The Natural Philosophy of Plant Form

AGNES ARBER

CAMBRIDGE
UNIVERSITY PRESS

CAMBRIDGE UNIVERSITY PRESS

Cambridge, New York, Melbourne, Madrid, Cape Town,
Singapore, São Paolo, Delhi, Mexico City

Published in the United States of America by Cambridge University Press, New York

www.cambridge.org
Information on this title: www.cambridge.org/9781108045056

© in this compilation Cambridge University Press 2012

This edition first published 1950
This digitally printed version 2012

ISBN 978-1-108-04505-6 Paperback

THE NATURAL PHILOSOPHY
OF PLANT FORM

By AGNES ARBER

HERBALS: THEIR ORIGIN AND EVOLUTION
A CHAPTER IN THE HISTORY OF BOTANY
(Second Edition)

THE GRAMINEAE: A STUDY OF CEREAL,
BAMBOO, AND GRASS

MONOCOTYLEDONS
(CAMBRIDGE BOTANICAL HANDBOOK)

WATER PLANTS

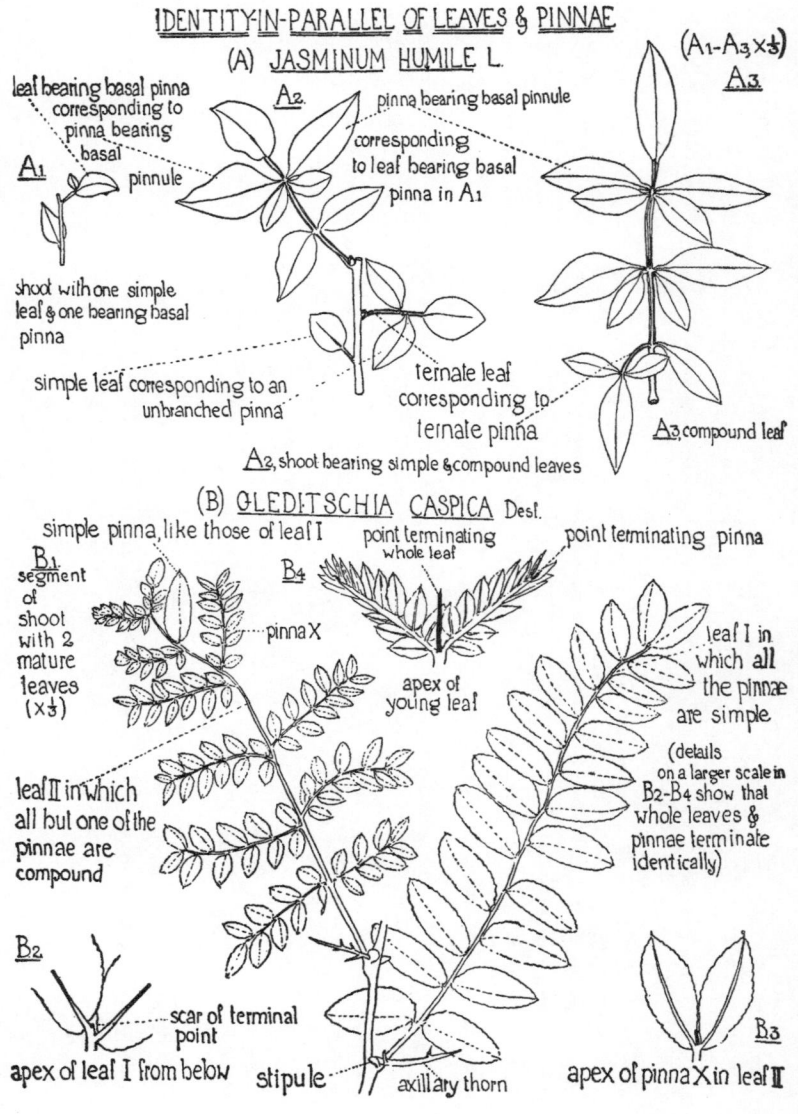

IDENTITY-IN-PARALLEL OF LEAVES & PINNAE
(A) JASMINUM HUMILE L.

$(A_1-A_3 \times \frac{1}{3})$
$\underline{A_3}$

$\underline{A_1}$

leaf bearing basal pinna corresponding to pinna bearing basal pinnule

$\underline{A_2}$.

pinna bearing basal pinnule

corresponding to leaf bearing basal pinna in A_1

shoot with one simple leaf & one bearing basal pinna

simple leaf corresponding to an unbranched pinna

ternate leaf corresponding to ternate pinna

A_3, compound leaf

$\underline{A_2}$, shoot bearing simple & compound leaves

(B) GLEDITSCHIA CASPICA Desf.

simple pinna, like those of leaf I

point terminating whole leaf

point terminating pinna

$\underline{B_1}$
segment of shoot with 2 mature leaves $(\times \frac{1}{3})$

pinna X

$\underline{B_4}$

apex of young leaf

leaf I in which all the pinnæ are simple

(details on a larger scale in B_2-B_4 show that whole leaves & pinnae terminate identically)

leaf II in which all but one of the pinnae are compound

$\underline{B_2}$

scar of terminal point

apex of leaf I from below   stipule

axillary thorn

$\underline{B_3}$

apex of pinna X in leaf II

THE
NATURAL PHILOSOPHY
OF PLANT FORM

BY

AGNES ARBER
M.A., D.Sc., F.R.S., F.L.S.

CAMBRIDGE
AT THE UNIVERSITY PRESS
1950

PUBLISHED BY
THE SYNDICS OF THE CAMBRIDGE UNIVERSITY PRESS
London Office: Bentley House, N.W. 1
American Branch: New York

Agents for Canada, India, and Pakistan: Macmillan

Printed in Great Britain at the University Press, Cambridge
(Brooke Crutchley, University Printer)

PREFACE

IN the present study I have tried to express certain general ideas, which have gradually disengaged themselves in my mind, in the course of a lifetime's concern with the morphology of flowering plants, both as it is understood to-day, and in its historical development from the time of Aristotle onwards. I began by thinking of this subject quite simply as a branch of natural science, but I have come finally to feel that it reaches its fullest reality in the region of natural philosophy, where it converges upon metaphysics, to which it brings its own, distinctively visual, contribution. In this book I have made a tentative and provisional attempt to review the relations of parts in the flowering plants in the light of those more universal, and also more stringent, modes of thought, which are characteristic of philosophy rather than of biology. There are indications that, when morphology is subjected to this discipline, its content may be unified by the synthesis of various theories that are, from the standpoint of analytical science, irreconcilable. The thread running through the following pages is thus a belief in the vital necessity of a linkage between morphological and philosophic thought.

This small treatise is the upshot of so many years that a *catalogue raisonné* of those to whom I have owed practical help or intellectual stimulus would expand this preface into an autobiography; so I must content myself with recording how deeply beholden I am to the generous comradeship of fellow-workers, above all when this has taken the form of enlightening criticism.

To my daughter, Muriel, I dedicate this book, in the consciousness of its having come into being on the background of our unending talks about "why things are, and that sort of thing".

AGNES ARBER

CAMBRIDGE
19 December 1949

ACKNOWLEDGEMENTS

I AM indebted to the Editor of *Biological Reviews* for permission to incorporate in this book parts of two articles of mine, which appeared in his journal. I have also to express my gratitude to the following publishers and editors, who have allowed me to quote from translations from the classics, for which they are responsible: the Clarendon Press, for certain passages from D'Arcy W. Thompson, *Historia animalium*, and W. Ogle, *De partibus animalium*—both in the Oxford translation of the Works of Aristotle; the Editors of the Loeb Classical Library, and Messrs W. Heinemann Ltd., for citations from W. S. Hett, *Aristotle on the Soul*, and A. Hort, *Theophrastus, Enquiry into Plants*; and Messrs Longmans Green and Co. Ltd., for an extract from W. Ogle, *Aristotle on Youth and Old Age*. In addition, I wish to thank Dr Robert E. Dengler, Professor of Classics, the Pennsylvania State College, for his kindness in letting me cite his version of the *De Causis Plantarum* of Theophrastus.

I desire also to acknowledge with gratitude the expert guidance and the invariable patience of the staff of the Cambridge University Press.

A. A.

CONTENTS

xi

CONTENTS

LIST OF ILLUSTRATIONS

xiii

LIST OF ILLUSTRATIONS

CHAPTER I

THE MEANING AND CONTENT OF PLANT MORPHOLOGY

IN these days of specialised study, the different branches of biology cannot but lead existences which are, to a great extent, isolated from one another. The aims which they pursue, and the highly technical methods by which these aims are achieved, differ so widely that one reminds oneself, with something of a shock, that all the branches are concerned with the same living world, and that their disjunction arises, not out of differences of content, but out of the divergent treatment which the mind accords to the same phenomena, when seen from various standpoints—"thinking makes it so". The different branches should not, indeed, be regarded as so many fragments which, pieced together, make up a mosaic called biology, but as so many microcosms, each of which, in its own individual way, reflects the macrocosm of the whole subject. The flowering plant, which in the present book will be our focal centre, offers innumerable 'microcosmic' aspects, varying according to the lines upon which it is considered; but we shall confine ourselves here to one chosen approach—that of morphology. This may seem a narrow road, but, rightly conceived, it should, like other biological paths, lead us towards infinite issues. By morphology we shall understand the study of *form*, giving this word, however, the wider connotation which it has, in general, lost; in modern speech it has become restricted, until it relates only to characters of superficial shape, while the adjective, 'formal', is reduced to mere triviality. A slight indication of the extent to which the word 'formal' has suffered degradation in the last few centuries is given by the phrase, "To make of him a formal man again";[1] for no one would now use this expression, as Shakespeare does, to include all the implications of a return to sanity. Some hint of

[1] *Comedy of Errors*, Act v, Sc. i. Moreover, in seventeenth-century philosophy, 'formal' may mean "that which has actuality or form"; cf. White, W. Hale, and Stirling, A. H. (1899), p. x.

the fuller meaning of 'form' is retained to-day in such collo-quialisms as "on the top of his form". In morphology our usage needs to be enlarged again until it can be brought into relation with that of Aristotle, to whom the scope of 'form' was wide enough to cover the whole of the intrinsic nature of which any given individual was a manifestation.

The term 'morphology' itself is also liable to be mulcted of its full measure of significance; in this connexion it is worth while to consider its etymology. The Greek μορφή as a rule means form, but Plato uses it in a more generalised fashion to denote a sensible character or quality.[1] Like the Latin *forma*, and like *makdome*, the corresponding old Scottish word, *morphe* sometimes carries the implication of elegance; that is to say, it conveys a certain suggestion of the harmony which is characteristic of the organisation of living things, and which St Augustine, who saw it in "the hearbes flower" and "the trees leafe", called "the peacefull concord of composition".[2] In analysing the term 'mor-phology', we have, furthermore, to consider the meaning of its termination, as well as of μορφή. We ought not to dismiss λόγος as equivalent merely to 'word'; it may stand for 'definition', 'explanation', 'that which is thought', and even for 'rational law',[3] or 'the formula giving the essence of a substance'. We may, indeed, understand morphology as involving the description and interpretation of the entire external and internal organisation of the plant, from the beginning to the end of its life-history, this organisation being viewed *sub specie formae*[4]—under the aspect of *form*—the fulness of content, with which Aristotle endowed it, being restored to this word. It is hence the business of morpho-logy to connect into one coherent whole all that may be held to belong to the intrinsic nature of a living being. Such a process clearly must transcend preoccupation with outward shape alone. There is indeed no justification for limiting morphology to ex-ternal features; as well as the outward form seen by the artist and systematist, it should invoke the analytic detail of anatomical and nuclear structure seen by the microscopist. Moreover, among

[1] Cornford, F. M. (1937), pp. 188, 199.
[2] Healey, J. (1610), *St Augustine, of the Citie of God*, bk v, chap. xi, p. 213; for a modern version of edition 2, see Healey, J. and Tasker, R. V. G. (1945), vol. I, p. 156.
[3] Whittaker, T. (1918), footnote, p. 36.   [4] See Arber, A. (1937a), p. 158.

plants, form may be held to include something corresponding to behaviour in the zoological field.[1] The animal can *do* things without inducing any essential change in its bodily structure. When a bird uses its beak to pick up food, the beak remains unchanged, but for most, though not for all plants, the only available forms of *action* are either growth, or discarding of parts, both of which involve a change in the size and form of the organism. Consider the growth phases of a bulbous plant. In the autumn it is a dormant bulb; in the spring and summer it puts forth roots, leaves and flowers, passes through a period of sex activity, and produces seeds; in the succeeding autumn, it loses its roots and the parts above ground, and returns to the bulb condition, sometimes accompanied or replaced by offspring bulbs. This sequence of growth stages, entailing shape transformations, corresponds to a whole series of motor acts[2] in an animal, and also to such directive activities as those concerned in sexual reproduction; assembling a store of food; and going into hibernation for the dead season. An example of this kind suggests that the contrast, which generally is assumed to exist between *form* and *function*,[3] has no reality when the word 'form' is given its full content. The treatment of the two conceptions as antithetic has, no doubt, been fostered by the neat alliteration of the phrase, but their assumed opposition is, in the main, traceable to the analogy, mistaken for something approaching an identity, between the works of man, and living beings themselves. In artificial constructions, the object which a man is making is first shaped, often in separate parts, and finally, when all is completed and fitted together, and some source of energy is supplied, the mechanism becomes capable of fulfilling the purpose for which its maker destined it. It is possible here to think of form and function as disjoined; but in living creatures there can be no such separation, for form (in the narrower sense), and function, are merely two aspects of the same unity. The word form, in its wider meaning, must be held to synthesise form in the more obvious sense, which is static, and function—the dynamic—which is the reverse side of the

[1] Russell, E. S. (1934), p. 89; the same idea seems to be hinted at in Crow, W. B. (1929), p. 30.

[2] For an account of a 'motor act' see Sherrington, C. (1940), pp. 205 *et seq.*

[3] The conceptions of form and function in zoology are treated on an historical basis in Russell, E. S. (1916).

same shield; in other words, form, as understood in morphology, should comprehend and fuse both static and dynamic elements.

The word 'form', as applied to plants, has indeed so far-reaching a connotation that it may induce a sense of hopelessness about the possibility of getting to grips with so extensive a notion. This is, however, a difficulty which resolves itself, since form, in its whole breadth of significance, finds a focus and expression in that aspect which is perceptible to our sight, supplemented by the sense of touch; for, in the "Figures, Fashions, and Shapes" of plants—to use Lyte's sixteenth-century wording[1]—all the elements belonging to form in its wider sense are made manifest, and brought into relation with our minds. Goethe long ago noticed that there was a tendency for scientific men to consider the external, visible, and tangible parts of living things as indications of their internal parts;[2] and modern work confirms the idea that form in the narrower sense often serves as an index to more recondite characters. It is well recognised, for instance, that the classificatory position assigned to a flowering plant, exclusively on the evidence of such outer morphological features as can be detected in herbarium material, frequently survives the test of subsequent detailed knowledge of the anatomy, life-history, chemistry, and ecology of the plant in question.[3]

We may sum up these considerations by saying that, to arrive at the fullest understanding of any individual plant form, we have, first, to realise it accurately by means of sensuous perception; secondly, to get the completest possible picture of it with the mind's eye—a picture which receives sculptural solidity from the data gathered by touch, and internal concreteness from knowledge of anatomical structure; and, thirdly, to advance beyond this representation, so as to grasp its underlying and surrounding context of significance, and to see it in its living aspect, and in its relation to other forms.

In the present book, evidence from abnormalities will sometimes be used to illustrate and supplement conclusions derived

[1] Lyte, H. (1578), title-page.
[2] Troll, W. (1926), p. 115 [Goethe, J. W. von, *Zur Morphologie. Die Absicht eingeleitet*].
[3] Cf. Diver, C. (1940), p. 305.

from the study of normal structures. This may seem to need justification, since evidence of this type is regarded by many botanists with distrust. It must be recalled that teratology can be viewed from two standpoints: firstly, as the study of abnormal forms, pursued for its own sake; and, secondly, as the same study pursued for the ulterior motive of discovering from it facts about ancestral history. This second aspect of teratology depends upon the assumption that clues to phylogeny are revealed in abnormal structures—an assumption which is both non-proven and improbable.[1] Biologists are thus rightly sceptical about the second aspect, but, unfortunately, this scepticism has often been allowed to extend to the first aspect, which, in justice, should not be placed under the same ban. Looking at the matter historically, we find that, before the Darwinian theory, with its phylogenetic corollaries, captured the imagination of botanists, the study of abnormalities was approached on much broader lines, as a help towards the understanding of normal forms. Early in the seventeenth century, Francis Bacon had written, concerning "the Errors of Nature, things strange and monstrous", that "he who knows her deviations will describe her ways with the greater accuracy".[2] Some two hundred years later, Goethe used the evidence of teratology to throw light upon normal processes,[3] while, early in the nineteenth century, Jäger[4] maintained that abnormalities in plants are subject to the same laws as those expressed in normal development. A little later de Candolle wrote of the regular order which lies hidden in abnormalities,[5] and Geoffroy Saint-Hilaire, in his book about monstrous forms, said that his main object in the study of teratology was to reach a deeper knowledge of the normal.[6] The attitude of these pre-Darwinian writers was determined by the truth—obvious but often overlooked—that macroscopic nature is never really anomalous, so that even the so-called 'abnormalities' are essentially law-abiding. This point was made explicit long ago by Montaigne, who recognised that nature could not be contravened—"rien n'est que selon elle, quel

[1] Cf. Arber, A. (1931 *a*), pp. 197–200.
[2] Bacon, F. (1620), lib. II, xxix, p. 241; translation in Kitchin, G. W. (1855), p. 184.
[3] Goethe, J. W. von (1790).   [4] Jäger, G. F. von (1814), p. 291.
[5] Candolle, A. P. de (1827), vol. II, p. 238.
[6] Geoffroy Saint-Hilaire, I. (1832–6), vol. I, p. 13.

qu'il soit";[1] while Sharrock, in the seventeenth century, in describing aberrant phyllotaxis, added, "Yet even in these, irregularities themselves, there often seems to be a greater curiousness, and most proper order."[2] Examples of such orderliness in disorder are indeed frequent; we may recall the parallelism among abnormalities often seen within groups of related plants (e.g. the Gramineae).[3] The existence of a regularity underlying the abnormal makes it possible to apply scientific method to the study of teratology, and to use the results in the interpretation of normal form. Abnormalities, like other exceptional cases, at least show incontestably, what the plant *can* do; it is thus, in its revelation of potentialities not usually actualised, that teratology may throw light upon normal happenings.

The modern outlook upon abnormalities represents, in some ways, a return to pre-Darwinian views. It is thus a cogent instance of the need for a nexus of historical ideas as a background to morphology. It is true that the historical study of botany, if treated superficially, is apt to degenerate into pretty and trivial antiquarianism, but if pursued as an exacting discipline, it bears directly upon current thought. Botany, in so far as it claims to be a branch of natural philosophy, can neglect its own history only at great loss to itself. In philosophy in general, it is part of the recognised task of present-day thinkers to consider, criticise, appraise, and re-appraise, the work of philosophers of the past, remote as well as near; such studies are regarded, not as contributions to history merely, but as an intrinsic part of living philosophy. This should, by rights, be true also of biology, which, like philosophy, *is* its own history; and a study of the course, which biological science has taken, confirms the idea that repeated scrutiny of its embryonic phases is essential to its progress. In the history of research and discovery, the further work which arises out of that of each pioneer, is, as a rule, concentrated in some one direction. Eventually all that can be gathered by pursuing that path becomes exhausted; but by this time the originator himself has been more or less forgotten, and the trail which he blazed is deserted in favour of routes newly opened

---

[1] Montaigne, Michel de (1906, etc.), livre II, chap. xxx, p. 515. This sentence is o.Montaigne's manuscript additions to the 1588 text.
[2] Sharrock, R. (1660), p. 145.   [3] Arber, A. (1934), pp. 385–98.

elsewhere. If, however, at this stage the pioneer's work were again examined, it might be found to contain the germs of other developments, which could equally well have come to full fruition, but which have never had the chance, because one offshoot, which was more completely native to its period, achieved a monopoly from the first. For this reason it is well to return, even at long last, to such early work as is notably rich in content, to see whether it still offers suggestions, which formerly passed unheeded because the time was not ripe for them, but which the intellectual climate would now foster. Originality is so rare in the human mind, that we need to harvest it to the last gleanings. In plant morphology, the case for a return to the renewed study of the pioneers is particularly strong. For this there are two reasons, arising out of the nature of the subject itself. One of these reasons is that morphological research, though it can make full use of the utmost refinements of technique, is yet not debarred from proceeding vigorously without any such aids; for even when naked-eye observation was the only channel through which information could be gathered, sound conclusions were reached by those gifted with the seeing eye, bodily and mental. The earlier workers were thus at less disadvantage in the study of plant form than in other botanical fields. The second reason accentuating the value, even to-day, of long-ago work in morphology, is that, being free from any fixed scheme of evolutionary pre-conceptions, the writers of the past were at liberty to concentrate on form *in itself*. This single-mindedness enabled them to go far, since it meant that their thought was not inhibited by doctrinaire attempts to force it to fit hypothetical history. The whole attitude of many post-Darwinian botanists, on the other hand, has been distorted, through trying to compel the study of form to subserve phylogenetic ends. The work of the Greeks shows us how far morphology was capable of advancing in the absence of modern technique, and without the rigid mental framework imposed by evolutionary theory. It is, indeed, difficult to imagine how any biologist, even with to-day's masse⟩ heritage of factual detail at his command, could better the broadly holistic view of the general nature of morphological thought set forth by Aristotle. He pointed out that, when any part or structure is under consideration, "it must not be supposed that it is its material composition, to which attention is being directed ... but the relation of each part to the total form. Similarly, the true object of architecture is not bricks, mortar, or timber, but the house; and so the principal object of natural philosophy is not the material elements, but the composite thing, and the totality of the form, independently of which these elements have no existence."[1] For the morphologist, Aristotle in truth remains, as for Dante long ago, "il maestro di color che sanno."[2]

[1] Ogle, W. (1912), vol. v [*De part. anim.* I. 5. 645a (Oxford trans.)], slightly modified after comparison with Peck, A. L. (1937).

[2] *Inferno*, canto IV, 131, "the Master of those that know."

CHAPTER II

THE PLANT MORPHOLOGY OF THE ARISTOTELIAN SCHOOL

MODERN developments in plant morphology cannot be understood without a study of the work of the Aristotelian school, since it is from this work that the whole subject took its rise. Writers on phytology have long recognised that no botanical book actually by Aristotle has come down to us;[1] Thomas Johnson, for instance, in 1633 quoted scholarly authority for "doubt of these bookes carried about in his name".[2] There are, it is true, allusions in Aristotle's authentic works to a treatise on plants, but it seems probable that these allusions relate to such parts of the writings of his follower, Theophrastus, as were set forth in the older philosopher's lifetime, rather than to a work of his own. We have, however, some knowledge of Aristotle's botanical views, since his biological principles are often applicable to the vegetable as well as the animal world, and, in the course of his zoological writings, he makes a number of references to the life and structure of plants. Special importance attaches to the distinction which he draws between the psyche[3] in plants, animals, and man. The plant has merely a 'nutritive' psyche, or vital principle; to this, in the animal, a 'sentient' psyche is added, while man has a 'reasoning' psyche, which includes the two lower grades, with an intellectual principle in addition.[4] In this view as to the relation of types of psyche in different classes of living things, Aristotle diverges from Plato, who attributed sensation to plants, though he denied them self-consciousness.[5] In man and animals, Aristotle locates the sensory psyche primarily in the heart. One of the reasons which led him

[1] For a detailed study of this subject see Senn, G. (1930).
[2] Johnson, T., *To the Reader*, in Gerard, J. (1633).
[3] It seems best to retain the Greek word, which is untranslatable; 'soul', which is commonly used for ψυχή, conveys a misleading impression.
[4] Lones, T. E. (1912), pp. 80, 95; Thompson, D'Arcy W. (1913), pp. 26–7.
[5] Cornford, F. M. (1937), pp. 302–3 [*Timaeus*, 76 E–77 C].

to that view was the central position of this organ,[1] and, when he turned to plants, he was induced by analogy to look for a corresponding central seat for the nutritive psyche. He found what he sought in the part of the seedling which forms the boundary between shoot and root. He writes: "that the main seat of the nutritive psyche is central is plainly to be seen both in plants and in animals. In plants it is manifested in the phenomena presented by the germination of seeds, and by grafts and cuttings. For invariably the development of a seed proceeds from a central spot. For all seeds consist of two valves, which are joined together at the point where the seed is connected with the mother-plant; and it is from this point as a centre, which belongs equally to either side, that both stem and root are given off when growth begins; for it lies midway between them and gives origin to both."[2] Aristotle's belief that the psyche of the plant resides in the cotyledonary node was thus based partly on the mistaken identification of this node with the point of attachment of the seed to the parent tissues (the hilum). He goes on to point out that each vegetative bud, also, may be a centre of root and branch development. In summarising Aristotle's views on the site of the plant psyche, one may easily make them appear more definite and rigid than they actually were; for we find that, though he believed he could locate the nutritive psyche specifically in seedlings and in buds, he also attributes the possibility of propagation from slips to the fact that "the vital principle exists potentially in every part of the plant".[3]

Aristotle opens his *Historia animalium* with a significant analysis of the living body; though this analysis was zoological in its first intention, it also has a bearing upon plant study. He tells us that "Of the parts of animals, some are simple: to wit, all such as divide into parts uniform with themselves, as flesh into flesh; others are composite, such as divide into parts not uniform with themselves, as, for instance, the hand does not divide into hands nor the face into faces." And he adds, "All those parts that do not subdivide into parts uniform with themselves are composed

[1] Ogle, W. (1912), vol. v [*Depart. anim.* II. 10. 656a; III. 4. 666a (Oxford trans.)].
[2] Ogle, W. (1897), pp. 64–5 [cap. III], translation slightly modified.
[3] Hett, W. S. (1935), pp. 402–3 [*On length of Life*, VI. 467a]; also pp. 50–1; 64–5; 76–7 [*On the Soul*, I. iv. 409a; I. v. 411b; II. ii. 413b].

of parts that do so subdivide, for instance, hand is composed of flesh, sinews and bones."[1] In another great zoological work, *De partibus animalium*, he carries the analysis a stage further back, and traces 'similar parts' (τὰ ὁμοιομερῆ) to certain elementary forces,[2] or primary substances,[3] from which they are compounded, while the 'dissimilar parts' (τὰ ἀνομοιομερῆ) are correspondingly compounded out of the similar parts. His description of the system of the living body is thus that it consists of elements, which are the material for τὰ ὁμοιομερῆ, which are, in their turn, material for τὰ ἀνομοιομερῆ, which, in the final synthesis, make up the organism as a whole. The division into similar and dissimilar—or homogeneous and heterogeneous—parts, is roughly equivalent to the more modern classification into tissues and organs. In the Oxford translation of *De partibus animalium*, it was noted that this equivalence was incomplete, because Aristotle included among homogeneous parts "much that we should not call tissue, e.g. the blood";[4] but in recent times there has been a return to Aristotle in this matter, for blood is now regarded as a liquid tissue.[5]

Aristotle held that the plant body was divisible on the same principles as that of animals, but, as it was only incidentally that he touched upon plants, there is more content for us in the writings of his successor, Theophrastus (370 B.C. to 285 B.C.), to whom it seems possible that he may have handed over, deliberately, the continuation of his work on the botanical aspect of biology.[6] Theophrastus was not more than fifteen years younger than Aristotle, and, like him, had studied under Plato, so that he is perhaps more justly described as a junior contemporary and colleague of Aristotle, than merely as his pupil.

The technical terms, which Theophrastus adopts for the minor parts of plants, which had not, like the principal parts, already acquired primary folk names, illustrate one of the more charac-

---

[1] Thompson, D'Arcy W. (1910), vol. IV [*Hist. animal.* I. 1. 486*a* (Oxford trans.)]. Thompson points out that Aristotle's distinction of similar and dissimilar parts is derived from Anaxagoras.

[2] Ogle, W. (1912), vol. V [*De part. anim.* II. 1. 1. 646*a* (Oxford trans.)].

[3] Peck, A. L. (1937), p. 12 [*De part. anim.*].

[4] Ogle, W. (1912), vol. V [*De part. anim.* note to II. 1. 646*a* (Oxford trans.)].

[5] Cf., for instance, Cabot, R. C. (1904), p. 4, and Roberts, M. (1926), p. 180.

[6] Cf. Strömberg, R. (1937), p. 25; for an important analysis of the relation which the biological work of Theophrastus bears to that of Aristotle, see Senn, G. (1933).

teristic features of the botany of the Aristotelian school—a constant reference to animals; in consequence, our botanical language, from classical times until to-day, has faithfully reflected zoological terminology. As examples from Theophrastus we may mention: σάρξ (flesh), for plant pulp; ῥάχις (backbone) for the midrib of a leaf; καρδία (heart) for heart-wood or the pith region; and ἴς (nerve or sinew) for a nerve in the leaf. The theory that the key to the understanding of plants was to be found in zoology, led Aristotle to suppose that the roots of plants correspond to the mouths of animals, since both roots and mouths are members by means of which food is absorbed; the plant, on this view, is comparable with an animal standing on its head.[1] This idea can be traced back to Democritus,[2] while Plato[3] speaks of man as like a plant whose roots are not in the earth, but in the heavens. The notion that the root corresponds to the mouth was handed on by Boethius, who, in the sixth century A.D., wrote of plants, "as it were, thrusting their head into the ground",[4] while, even in the sixteen hundreds, Bacon quotes with approval, "Homo est planta inversa."[5] As late as the mid-nineteenth century, a well-known text-book[6] sponsored the crudely conceived analogy between root and mouth. The author writes that "many insects support themselves wholly by suction; and...all plants do the same". He goes on to explain that plants are provided, "not with a single sucker, like the leech or the flea, but with many". These suckers he describes as taking the form of spongelets at the extreme tips of the rootlets, which absorb 'carbonised water'.

Theophrastus, with his wider botanical knowledge, tended to be more cautious than Aristotle about analogies between the animal and vegetable worlds: he says that "we must not assume that in all respects there is complete correspondence between plants and animals";[7] but with this reservation, he often turns to

---

[1] Peck, A. L. (1937), pp. 370–1 [*De part. anim.* IV. x. 686*b*]; Hett, W. S. (1935), pp. 69, 89 [*On the Soul*, II. i. 412*b*; II. iv. 416*c*].

[2] Cornford, F. M. (1937), p. 357.

[3] Cornford, F. M. (1937), p. 353 [*Timaeus*, 90].

[4] Boethius, A. M. T. S. (1609), bk. II, xi, opp. p. 80; for a revised version of this translation, see Stewart, H. F. and Rand, E. K. (1918), *The Consolation of Philosophy*, pp. 280–1.

[5] Bacon, F. (1631), p. 151; this edition quoted as the first is not accessible to the writer.

[6] Rennie, J. (1849). pp. 88, 92.   [7] Hort, A. (1916), vol. I, pp. 6, 7 [I. i. 4].

analogy for help, and he is indeed right in doing so. For human thought can proceed only by passing from the known to the unknown, on the assumption that there is some degree of real analogy between the two.[1] It has been a certain handicap to the study of plants, that, owing to man's primary interest in the animal world, to which he belongs, his attention was naturally first focused upon this, and plants were thus left to be interpreted subsequently through zoological comparisons; the reverse process might at times have been more illuminating.

The botanical work of Theophrastus is of such importance, as the basis on which even to-day we still build, that it is necessary to consider it in some detail, at least as regards those aspects which bear definitely upon morphology. Our knowledge of his output is derived from two books, *De causis plantarum* (περὶ φυτῶν αἰτιῶν), and *De historia plantarum* (περὶ φυτῶν ἱστορία). It has been suggested[2] on internal evidence that *De causis* is relatively early, while the *Historia*, in which the vocabulary shows greater richness, precision, and specialisation, represents a group of later works. The influence of Aristotle is conspicuous in *De causis*, while the *Historia* shows more independence.[3] Both treatises have some of the characteristics of lecture notes.[4] The fact that neither of them is a fully integrated individual work, representing a final written version of the author's views, does not add to the ease of interpretation; but, even so, a great part of their content is lucid and of profound interest.

The *Historia* opens with an account of the general scope of botany. Theophrastus tells us that we must take into considera-tion: (1) the parts of plants; (2) their qualities; (3) the way in which their life originates; and (4) the course which their life follows. He adds that behaviour and activities, such as we witness in animals, are not to be found in plants. This classification, which is remarkable for its comprehensiveness, might be restated in modern terms as representing the division into (1) morphology; (2) physiology and biochemistry; (3) the study of reproduction

[1] Cf. Hort, A. (1916), vol. I, pp. 18, 19 [1. ii. 4]; on analogy in the history of science, see Arber, A. (1946 b).

[2] Strömberg, R. (1937), p. 136; for a careful chronological study of the botany of Theophrastus, see Senn, G. (1933).

[3] Senn, G. (1930), p. 113.

[4] 'Vorlesungsmanuskripten', Strömberg, R. (1937), pp. 69–70.

and development; (4) the study of life-histories. Theophrastus goes on to say that, of these subjects, that dealing with the parts of plants (morphology) presents the greatest difficulty. Writing, as he did, more than 2000 years ago, he was naturally not in a position to make a critical comparison between the obstacles to be overcome in these various regions of research, but his appreciation of the complexity of morphology indicates that this was the field into which he had the clearest insight; it is indeed the field in which his influence has been most lasting, while that of Aristotle has been more deeply felt in the study of behaviour, and of 'purposeful' structure.[1] The difference of emphasis in the work of these two pioneers possibly represents a natural cleavage between the standpoint of the botanist and zoologist—the botanist inclining to the consideration of form, and the zoologist to that of function.

We have already mentioned Aristotle's realisation of the existence in the animal body of homogeneous parts (tissues), which are the components of heterogeneous parts (organs). Among plants, correspondingly, he recognised organs, but he says that they are of a simpler type; he instances leaf and seed-vessel.[2] Theophrastus carries the analysis to a further point, and classifies plant organs under two categories: *the main parts*, such as root, stem, bough, and twig; and the *annual parts*, such as flower, leaf, fruit, and 'new shoot'. On comparing the parts of plants with those of animals, he is struck by the impermanence of the plant members of the second category, and with the fact that trees make fresh shoots every year, so that the number of parts is indeterminate and continually changing, whereas in the higher animals the number of organs is fixed and definite. He evidently feels that this is a divergence which must be accepted as such, without trying to force our conception of the plant into the framework derived from zoology. He holds that "it is waste of time to take great pains to make comparisons where that is impossible";[3] this may seem to be a glimpse of the obvious, but in those days the overstrained animal analogy had a stranglehold which it is hard now to realise.

---

[1] Cf. Thompson, D'Arcy W. (1913).
[2] Hett, W. S. (1935), p. 69 [*On the Soul*, ii. i. 412b].
[3] Hort, A. (1916), vol. i, pp. 6, 7 [i. i. 4].

As the root comes first in Theophrastus's enumeration of the main parts of plants, we will begin with this member in considering his views in more detail. By the botanists who preceded him, every part of the plant which carried on its existence under the earth, appears to have been called a root,[1] but Theophrastus, though with characteristic caution, challenges this as a general assumption. He shows that it would involve describing as roots, the stalks of 'long onion,' and also such underground plants as the truffle. He points out, moreover, that the 'roots' [rhizomes] of reeds and of dog's-tooth grass are jointed and resemble the above-ground parts. Again, in discussing the solid 'head' [bulb] of squill or onion, he decides that this is one kind of 'root', but that roots of a second kind grow out of it, and that it is these which resemble ordinary roots. He is, indeed, almost disposed to discard the term 'root' for the bulb, on the ground that a true root tapers continuously to its tip, whereas the form of the 'head' is quite different. Apart from such morphological features, he felt that it was potentialities, rather than situation, which should be emphasised in the definition of the root; he recognised, for instance, as an essential, though negative character, that it does not produce leaves. This comes out in his admirable description of the banyan (*Ficus benghalensis* L.), in the course of which he tells us that it sends down roots every year from the older branches, which "make, as it were, a fence about the tree, so that it becomes like a tent, in which men sometimes even live. The roots as they grow are easily distinguished from the branches, being whiter, hairy, crooked, and leafless".[2] It should be noted, however, that ἄφυλλοι, 'leafless', is a sixteenth-century emendation due to d'Aléchamps,[3] which may be invalid; uncertainties such as this are unfortunately frequent in the *Historia*, since even the best manuscripts are corrupt.

In considering trees, Theophrastus distinguishes the tap root, noting that, in the firs there is one long root, "which runs deep, and a number of small ones branching from this", whereas in the olive the tap root is small, "while the others are larger and,

---

[1] Strömberg, R. (1937), p. 58.

[2] Adapted from Hort, A. (1916), vol. I, pp. 312–15 [IV. iv. 4].

[3] Hort, A. (1916), vol. I, footnote 2, p. 314.

as it were, spread out crabwise ".[1] It was characteristic of the early botanists that they took a particular interest in roots, partly because that region of the plant was specially valued for medicinal purposes. Theophrastus, for this or other reasons, paid attention to root systems, and it is held by Strömberg, to whom we owe an exhaustive study of his work, that his writings, even to-day, are among the best and most thorough on the root morphology of Mediterranean trees.[2]

Theophrastus distinguished between adventitious and lateral roots, though without giving them different names.[3] As an example of his insight into another distinction, we may recall that he not only described what we now know to be the assimilating roots of the water-chestnut (*Trapa natans* L.), but he advanced so far towards their identification as to see that they partook neither of leaf nor of stem character.[4]

Next to the root, in Theophrastus's enumeration, come the trunk, bough, and twig, that is to say, the parts which we now think of together under the name of stem. He showed appreciation of a fundamental feature of shoot structure, when he recognised the jointed character of the reed stem, and pointed to sedge and rush as being smooth and jointless; he is thinking only, of course, of the aerial axes. It is thus evident that he distinguished nodes and internodes; indeed, when we speak of an 'internode', we are merely using the Latin translation of the term, μεσογονάτιον, which he invented.

Theophrastus was interested not only in the external form of plants, but also in their anatomy, which, as he points out, corresponds to the study of animals by dissection; this comparison does not seem to have been revived until the mid-seventeenth century, when Francis Glisson re-enunciated it.[5] In plants Theophrastus distinguishes bark, wood, and core—the latter term including both heart-wood and pith. One of his best detailed anatomical observations relates to the structure of the palm stem. He records that "the fibres [vascular bundles] do not run

---

[1] Hort, A. (1916), vol. I, pp. 40–1 [I. vi. 3].
[2] Strömberg, R. (1937), p. 77.
[3] Cf. descriptions of basil and blite in Hort, A. (1916), vol. II, pp. 73 [VII. ii. 7].
[4] Hort, A. (1916), vol. I, pp. 358–9 [IV. ix. 3].
[5] Arber, A. (1941 b), pp. 220–1.

throughout the wood, nor do they run to a good length, nor are they all set symmetrically, but run in every direction".[1]

After the permanent organs—root and stem—Theophrastus discusses the impermanent members; we may take for first consideration his treatment of the leaf. In those days nothing was known of its major functions in the life of the plant, and it was regarded merely as an accessory to fruit production.[2] Leaves seem, indeed, to have attracted comparatively little general attention in classical times, for, although there are several ancient Greek words for branch and for fruit, we find that the leaf is poor in special names. It is possible that the small-leaved character, frequent in the Mediterranean flora, may have hindered appreciation of the importance of foliage.[3]

Owing to the stress which Theophrastus laid upon the distinction between the permanent and transitory parts, he was much intrigued by leaf-fall. He understood that there was no hard-and-fast difference between evergreen and deciduous types. He says that in evergreen trees "the shedding and withering of leaves take place by degrees; for it is not the same leaves which always persist, but fresh ones are growing, while the old ones wither away".[4] The search for an animal analogy does not yield him anything more enlightening than the comparison of the fall of the leaf with the shedding of a stag's antlers.

When he turns to the form of leaves, Theophrastus describes the sessile and petiolate; the various shapes which the blade assumes; and the characters of margin and apex. As an example of his method we may quote his account of the leaves of the hop-hornbeam (*Ostrya carpinifolia* Scop.). They are, he says, "in shape like those of a pear, except that they are much longer, come to a sharp point, are wider, and have many nerves, which branch out like ribs from a strong straight one in the middle, and are thick; also the leaves are wrinkled along the nerves, and have a finely serrated edge".[5] His account of the venation of this dicotyledon may be contrasted with what he says of a monocotyledon, smilax, in which he alludes to the curved longitudinal

[1] Hort, A. (1916), vol. I, pp. 436, 437 [v. iii. 6].
[2] Hort, A. (1916), vol. I, pp. 16, 17 [I. ii. 1].
[3] Cf. Strömberg, R. (1937), p. 143.
[4] Hort, A. (1916), vol. I, pp. 666–7 [I. ix. 7].
[5] Modified from Hort, A. (1916), vol. I, pp. 224, 225 [IX. x. 3].

veins which do not start from the midrib. Another feature, which he notes for a monocotyledonous leaf, is the hollowness of that of the onion. The subject of succulence in leaves interested him, and he distinguishes those that, though succulent, are dorsiventral, thus showing "their fleshy character... in the flat instead of in the round".[1] He is particularly clear about the dorsiventrality of leaves in general, noting that the upper surface is greener and smother, while the veins show on the lower surface. Among the most striking of Theophrastus's achievements is his detection of the equivalence between the single simple leaf, and a compound leaf, the whole of which may, he suggests, be considered as one leaf, in such a tree as the ash, because it is shed all at once.[2] His account of the foliage of the service-tree (*Pyrus Sorbus* Gaertn.) is a specially good example of his treatment of the compound leaf. Incidentally it also illustrates the essentially *comparative* style of his leaf descriptions—a feature which still survives in a number of specific names, referring to leaf characters, which we use to-day, such as *salicifolia, betulifolia,* etc.[3] Theophrastus writes that the leaves of the service-tree "grow attached to a long fibrous stalk, and project on each side in a row like the feathers of a bird's wing, the whole forming a single leaf but being divided into lobes with divisions which extend to the rib; but each pair are some distance apart, and when the leaves fall, these divisions do not drop separately, but the whole wing-like structure drops at once... at the end of the leaf-stalk there is an extra leaflet, so that the total number of leaflets is an odd number. In form the leaflets resemble the leaves of the 'fine-leaved' bay, except that they are jagged and shorter and do not narrow to a sharp point but to a more rounded end."[4]

Theophrastus had an eye for certain anomalous types of foliage; he mentions, for instance, as a peculiar case, that butchers' broom (*Ruscus aculeatus* L.) and Alexandrian laurel (*R. Hypophyllum* L.) bear fruit "on the midrib of the leaf".[5]

It is noteworthy that, in enumerating the annual parts, Theophrastus includes the 'new shoot', thus putting it into the

[1] Hort, A. (1916), vol. I, pp. 70, 71 [I. x. 1].
[2] Hort, A. (1916), vol. I, pp. 230, 231 [III. xi. 3].
[3] Greene, E. L. (1909), p. 105.
[4] Hort, A. (1916), vol. I, pp. 240, 241 [III. xii. 7].
[5] Hort, A. (1916), vol. I, pp. 266, 267 [III. xvii. 4].

same category as leaf, flower, etc. It would be incorrect to suppose from this that he consistently regarded the shoot as, in some sense, a morphological unit, but both he and Aristotle seem to have had at least a glimmering of this conception. Aristotle says that "plants are always being reborn; that is why they last so long. For some branches are always new, while others grow old."[1]

The ideas held by Theophrastus on shoot morphology in general were surprisingly enlightened. From certain of his descriptions it emerges that he recognised the difference between monopodial and sympodial growth. This is indicated, for instance, in his account of galingale (*Cyperus longus* L.), though he does not, of course, use the terms of modern botany, which, in the following citation, have been added in square brackets. "This plant", he says, "is peculiar in its way of shooting and originating; for from the trunk-like stock [mature rhizome-segment belonging to the preceding year] grows another slender root sideways [young rhizome-segment belonging to the current year], and on this again forms the fleshy part which contains the shoot from which the stalk springs [the terminal bud which grows up to form the erect reproductive shoot]."[2] He also points out that in the majority of trees "the growth is from the top of the shoots and also from the side-buds"; but "in some cases the growth is not from the top, but only from the side-buds". He adds that "in certain trees the buds end in a single leaf; wherefore it is reasonable that they should not make fresh buds and growth from this point, as they have no point of departure".[3] Despite its obscurities, it seems natural to take this description as hinting at the process of replacement of the main line of growth by a lateral line. Another illustration of the attention which Theophrastus paid to shoot development, is that he noticed the tillering of cereals—"new growth the next year from plants which are roughly treated or trodden down so that hardly anything remains visible, as happens when an army has marched over the field".[4]

---

[1] Hett, W. S. (1935), pp. 400–1 [*On Length of Life*, VI. 467 *a*].
[2] Hort, A. (1916, vol. I, pp. 364, 365 [IV. x. 5], except the words in square brackets.
[3] Hort, A. (1916), vol. I, pp. 192, 193 [III. vi. 2, 3].
[4] Hort, A. (1916), vol II, pp. 186–9 [VIII. vii. 5].

Theophrastus had a unified conception of the whole life-history, since he accepted from Aristotle [1] the idea that plant growth "leads up to reproduction as the completion of the process".[2] He realised, also, that reproduction, paradoxically, represents, not only the culmination of growth, but also its inhibition. "It happens", he says, "that, when the trees leaf very luxuriantly, they are more likely to be sterile, and whenever they bear copiously, they are more likely to leaf poorly, as though Nature could not satisfy both elements of the tree's growth but must spend her resources first on one and then on the other."[3] Since Theophrastus knew nothing of the actual process of fertilisation, as revealed by microscopic study, he does not distinguish, as a modern botanist does, between sexual and vegetative reproduction. He suggests that a bulb is, so to speak, "an embryo or fruit; wherefore those who call such plants 'plants which reproduce themselves underground' give a fair account of them".[4]

By the term flower (ἄνθος) Theophrastus appears to mean the corolla primarily, but also the stamens and styles. He describes flowers with free petals (e.g. those of almond, apple, and leguminous plants) as 'leafy' (φυλλώδη); when the petals, as in the rose and lily, surround a very conspicuous apparatus of stamens and styles, he calls the flower 'twofold', since he holds that there is 'another flower inside the flower'. He considers that the corolla of the bindweed (*Calystegia sepium* R. Br.), and the lower part of the narcissus perianth, may each be regarded as a single leaf, but he expresses this in a way suggesting that he, in fact, understands that the 'single' leaf is actually complex.[5] He observed that, in the apple and rose, "the flower is on the top of the fruit-case",[6] and he finds that the situation is similar in thistle, cucumber, etc. He expresses some doubt as to whether the term 'flower' should be extended to cover such developments as the catkin of the nut; the 'oak-moss' [male catkin] of the oak; and the 'flowering tuft' [male cone] of the pine. Nevertheless he gives an admirable account of the male catkin of the filbert (*Corylus Avellana* L.),

1 Hett, W. S. (1935), pp. 86, 87 [*On the Soul*, II. iv. 415*b*].
2 Hort, A. (1916), vol. I, pp. 6, 7 [I. i. 3].
3 Dengler, R. E. (1927), p. 125, [I. xx].
4 Hort, A. (1916), vol. I, pp. 46, 47 [I. vi. 9].
5 Hort, A. (1916), vol. I, pp. 90, 91 [I. xiii. 2].
6 Hort, A. (1916), vol. I, pp. 92, 93 [I. xiii. 3].

which he calls its clustering growth, produced after the casting of the fruit; he describes it as consisting of several objects, as large as a good-sized grub, growing from one stalk. "Each of these", he says, "is made up of small processes arranged like scales, and resembles the cone of the fir, so that its appearance is not unlike that of a young green fir-cone, except that it is longer and almost of uniform thickness from end to end. This grows through the winter; when spring comes the scale-like processes open and turn yellow; it grows to the length of three fingers, but, when in spring the leaves are shooting, it falls off." [1] The botanist of to-day, who watches through the winter months for the maturing of the hazel catkins, cannot but feel a delicate link of sympathy with Theophrastus, who must have gazed at them with the same sense of hopefulness, in the Lyceum garden, more than two thousand years ago.

It was impossible at that date for Theophrastus to understand the exact nature of the female part of the flower, but he sometimes identifies and describes it aright. He says, for instance, that in the 'Median' or 'Persian apple' (*Citrus* sp.), those flowers "which have, as it were, a distaff projecting in the middle are fertile". [2] The best word picture which he gives, including flowers and fruit, as well as vegetative organs, is perhaps that of the sacred lotus (*Nelumbium speciosum* Willd.); in the sixteenth century Andrea Cesalpino singled out this description as being most beautifully done. [3] It is worth citing, not only for its quality, but also since it illustrates Theophrastus's view that habitat and internal structure are integral elements in the plant's characterisation; it is, more-over, a good example of his emphasis upon comparison. He tells us that "the 'Egyptian bean' grows in the marshes and lakes; the length of its stalk is at longest four cubits, it is as thick as a man's finger, and resembles a pliant reed without joints. Inside it has tubes which run distinct from one another right through, like a honey-comb: on this is set the 'head', which is like a round wasps' nest, and in each of the cells is a 'bean', which slightly projects from it; at most there are thirty of these. The flower is twice as large as a poppy's, and the colour is like a rose, of a deep

[1] Modified from Hort, A. (1916), vol. I, pp. 190, 191 [III. v. 5, 6].
[2] Hort, A. (1916), vol. I, pp. 312, 313 [IV. iv. 3].
[3] Cesalpino, A. (1583), lib. xv, cap. VIII, p. 569.

21

shade; the 'head' is above the water. Large leaves grow at the side of each plant, equal in size to a Thessalian hat; these have a stalk exactly like that of the plant. . . . The root is thicker than the thickest reed, and is made up of distinct tubes, like the stalk."[1]

To Theophrastus the word seed (σπέρμα) meant—as it generally does to-day to practical men—any one-seeded unit, rather than merely a seed, in the strict sense in which this word is used by the modern botanist. Indeed, with no magnifying glass, and no clear notions about the nature of small flowers, it was often difficult for him to recognise the seed at all; he says, for instance, that in thyme "it is not easy to find, being somehow mixed up with the flower".[2] One realises how far he would have advanced in detailed knowledge, if he had enjoyed any of the facilities of modern technique, when one sees how clearly he was able to distinguish the seed from surrounding structures in the cypress cone, which, as it is on a relatively large scale, can be studied, up to a certain point, with the naked eye. He notes that "it is not the entire globular fruit of this tree that is the seed, but the light and, as it were, bran-like slight object which grows in the midst of the fruit, and which flutters away when the fruit breaks open".[3]

Theophrastus was aware of the features which we now call dicotyledony and monocotyledony of the embryo, though he did not arrive at any analysis of them. He mentions that "In almond, hazel, acorn. . . the bud [plumule] first begins to grow within the seed itself, and as it increases in size the seeds split—for all such seeds are in a manner in two halves, and those of the leguminous plants again all plainly have two valves and are double. . . ; but in cereals. . . the seeds are in one piece."[4]

A natural result of ignorance of sexuality in plants, was that seeding was not, at the time of Theophrastus, regarded, as it is now, as an almost universal occurrence. He thus thinks it necessary to give evidence for supposing that some trees, e.g. the willow, which appear to him to have no fruits, probably reproduce

[1] Hort, A. (1916), vol. ɪ, pp. 350–3 [ɪv. viii. 7].
[2] Hort, A. (1916), vol. ɪɪ, pp. 8, 9 [vɪ. ii. 3].
[3] Dengler, R. E. (1927), p. 33 [ɪ. v].
[4] Hort, A. (1916), vol. ɪɪ, pp. 150, 151 [vɪɪɪ. ii. 2].

themselves by seed, and adds that this is true of certain herbaceous plants, such as thyme. "As for the plane," he says, "it obviously has seeds, and seedlings grow from them. This is evident in various ways, and here is a very strong proof—a plane-tree has before now been seen which came up in a brass pot."[1]

Whatever the defects of Theophrastus's knowledge of the seed—defects due to failures in detailed observation—his broad general conception, as expressed in *De causis plantarum*, could scarcely be improved. He tells us that "all plant seed has in itself a certain amount of nourishment which is produced with it at the beginning, just as is the case with eggs. Wherefore the remark of Empedocles is not badly put when he says: 'And the tall trees oviposit'; for the development of seeds is very much like that of eggs except that we must extend Empedocles's dictum to include the seeds not only of trees but of all plants, for every kind of seed has nourishment in itself. Seeds are therefore capable of lasting for some time and are not like those animals' eggs that perish at once when separated from the parent."[2]

So cursory a sketch as we have attempted here, cannot do full justice to the work of Theophrastus; it can only serve to indicate the qualities which have earned it its fundamental position in plant morphology. In the next chapter we will touch briefly upon some of its sequels.

[1] Hort, A. (1916), vol. I, pp. 160–3 [III. i. 2].
[2] Dengler, R. E. (1927), p. 45 [I. vii.].

CHAPTER III

THE PLANT MORPHOLOGY OF
ALBERTUS MAGNUS AND
ANDREA CESALPINO

IT is one of the most salient features in the history of biology that the theory of botany, set forth by the Aristotelian school, kept its pre-eminence, without essential criticism or addition, for some 2000 years. In the present outline we will choose only three of its exponents to represent this long period—Albertus Magnus in the Middle Ages; Andrea Cesalpino in the late renaissance; and Joachim Jung in the seventeenth century. By the time we reach such younger contemporaries of Jung's as Malpighi and Grew, classical influences, though still strongly felt, are less completely dominant.

Albertus Magnus (*d.* A.D. 1280) was a great scholastic philosopher, who, in the present century, has been canonised. He had an original mind, and his work revitalised the botany of the Greeks. He had no access to the *Historia* and *De causis* of Theophrastus, and his only source for Aristotelian botany appears to have been a small treatise, called *De plantis*,[1] now attributed to Nicolaus Damascenus,[2] who flourished at the beginning of the Christian era. It was derived chiefly from Theophrastus, and to a small extent from Aristotle. *De plantis* had a strange history: the Greek of Nicolaus was translated into Syriac; thence into Arabic; and thence into the Latin version which Albertus used, and which he revered as representing a genuine treatise by Aristotle; and finally into Greek again, in which guise it has taken its place in editions of the works of Aristotle.

The book, *De vegetabilibus*,[3] in which Albertus commented on

[1] Forster, E. S. (1913), and Hett, W. S. (1936).
[2] On Nicolaus Damascenus and his work, see Meyer, E. H. F. (1854–7), vol. I, pp. 324–33; vol. IV, p. 38.
[3] For text see Meyer, E. H. F. and Jessen, K. F. W. (1867); and for commentary, including an analysis of which use has been made here, Meyer, E. H. F. (1854–7), vol. IV, 1857; see also Sprague, T. A. (1933 *a*, *b*).

24

'Aristotle', and incidentally put forward his own individual views on botany, deserves detailed study, but we can only here allude to a few points in which he diverges from Theophrastus, or advances further upon the same lines. The way in which Albertus analyses the make-up of the plant is somewhat modified from that of his predecessor. Albertus distinguishes, firstly, *integral essential* parts, serving for the maintenance of the individual. One of these, which he regards as containing the potentiality of all the others, he calls the sap (potentiâ quaelibet pars plantae quod vocatur succus). It is perhaps not unreasonable to regard Albertus's idea of 'sap', as equivalent to the nineteenth-century conception of protoplasm. Some notion of the kind must have been in the mind of Theophrastus also, for he says, "from one stuff and under the influence of one single generating cause do the parts come".[1] Besides the 'sap', which represents the parts *in posse*, Albertus counts among his integral essential parts those which actually exist, *in esse*. This group (Partes...quae actu sunt partes plantae) consists of members, such as root, stem, branch, etc. On the Aristotelian principle, he regards them as constructed out of 'similar parts', such as wood, and plant pulp (caro herbalis). He distinguishes a second set of 'essential parts', as being not integral to the individual; he calls them 'accidental', because they do not last throughout the life-time of the plant, but he holds them to be 'essential', because they serve for reproduction; these parts are leaves, flowers, fruits, and seeds. The inclusion of leaves in this list seems surprising to us to-day, but it was consistent with the view of the Aristotelian school, that the function of foliage was to screen the fruit. Albertus completes his analysis by referring to 'accidental non-essential parts', such as spines.

In his estimate of the relation of plant members, Albertus assigns the primary position to the root; this was a logical result of ignorance of the physiological activities of the leaves. He compares the life of the plant to that of a household, in which the root represents the paterfamilias, who is the one source of supply to a group of sons and servants, who all have their special roles to play. It is not only from the standpoint of function, but also from that of structure that Albertus regards the root as the domi-nating organ. He thinks that the root both has the free subter-

[1] Dengler, R. E. (1927), p. 73 [1. xii].

ranean existence, which is universally recognised, and is also continued into the above-ground region of the plant in the guise of pith, thus forming a 'core' to the whole organism. He says that the pith of the shoot represents the root, as the spinal cord in animals represents the brain (Medullae autem plantarum videntur esse vicarii virtutis radicalis, sicut nucha in animalibus est vicarius cerebri). This is perhaps a development of Plato's view that in animals the marrow—of which the brain was the principal development—was the life-substance.[1]

Much as Albertus was influenced by animal analogies, he recognised the necessity for caution in their use. He points out, for instance, that the so-called 'veins' in plants which convey the sap, such as the threads which can be pulled out when the plantain leaf is snapped across, do not really correspond to animal veins.

One of the most acute of Albertus's botanical observations concerns the relation of thorns and spines. He does not use different terms for these two structures, but, in fact, he distinguishes them sharply, recognising that thorns are productions of the wood, while spines are superficial (Sunt enim spinae, quae ex profundo plantae educuntur;...Sunt etiam quaedam spinae, quarum basis non profundatur in corpus plantae sed stat super corticem, quasi extrinsecus adhaerens plantae). Considering how clearly Albertus understood many morphological features, it is surprising to find how ready he was to assign to them physical explanations of a wholly baseless type; but it is, of course, natural that he should have been unaware that the physics and chemistry of his day were powerless to assist him in interpreting what he saw. His account of the reasons for the form of spines may serve as an example of these speculations as to causes, which he put forward with as much assurance as if they were observed facts. "Certainly", he writes, "the cause [of spine-form] is that the material, by the natural power of the plant, is first all expelled to the surface, and, on account of its viscosity, it cannot achieve a pointed shape through the heat of the plant alone. But when it reaches the surface, the inner heat of the material is helped by the power of the sun, and then it becomes acutely sharpened; and because the material is viscous, heat and viscosity begin to pull

[1] Cornford, F. M. (1937), pp. 293–5 [*Timaeus*, 73 B–D].

in contrary ways, the heat, indeed, tending to extension, and the viscosity to downward curvature, and for this reason the spines are like hooks, based upon the outer covering of the plant." It is to be supposed that such an 'explanation' satisfied Albertus's desire to discover efficient causes for the phenomena which he observed. Exactly the same attitude recurs, even in the seventeenth century, in the writings of Joachim Jung and Nehemiah Grew, in which penetrating morphological insight is combined with the wildest suggestions as to the physical and chemical factors involved.

One of the points, in which Albertus shows some advance upon Theophrastus, is in a more methodical analysis of the forms of flowers. He recognised alternating whorls of floral parts, and, in the lily, he distinguished the style from the stigmatic region, and the filament from the anther.[1] An odd lapse, which perhaps has some significance, is that he speaks of the flower of the columbine (*Aquilegia*) as resembling *four* eagles. This sort of mistake reappears in the work of other early botanists. In the Vienna manuscript of Dioscorides (*c.* A.D. 512), the very beautiful, and otherwise accurate, drawings of the pimpernels (*Anagallis*) represent the flowers as 4- instead of 5-petalled.[2] Moreover, at a much later date, Francis Bacon wrote of "Lillies, Flower-de-Luces, Borage, Buglosse", as 4-petalled flowers.[3] Bacon had himself previously complained that nothing in natural history was, in his time, numbered and measured.[4] One must either suppose that he himself was one of those whom he arraigns, or else—since the statement about petal numbers occurs in a posthumous work—it may be fairer to hold that he, personally, was not responsible. If any stress can be laid upon these scattered examples, in which parts in threes or fives are miscounted as fours, they may perhaps be held to suggest that the realisation of the importance of the numerical element in the study of living creatures is of relatively modern growth.

[1] Sprague, T. A. (1933*a*), p. 437.
[2] Black and white sketches of these drawings are reproduced in Gunther, R. T. (1934), pp. 223, 224.
[3] Bacon, F. (1631), cent. VI, p. 144; quoted from the third edition, since the first (1627) is not accessible to the writer.
[4] Bacon, F. (1620), lib. I. xcviii, p. 117, "Nil debitis modis exquisitum, nil verificatum, nil numeratum, nil appensum, nil dimensum in Naturali Historiâ reperitur"; translation in Kitchin, G. W. (1855), p. 80.

After Albertus, the next great figure, which we have taken to represent the classical tradition, is Andrea Cesalpino (1519–1603). He was both a physician, and a devoted student of Aristotle, whose philosophy he interpreted in a sufficiently original way to mark him, for some critics, as a precursor of Spinoza.[1] Aristotelian thought had so largely formed his mind, that it is not surprising that, though his botanical book, *De plantis* (Florence, 1583), dates from some three hundred years later than the writings of Albertus, he yet adopted the Theophrastean scheme more wholeheartedly than his medieval predecessor. To Cesalpino, an Italian, the classical way of thought must have been more completely native and compelling than it was to such men as Albertus, who were, to some extent, screened from the more direct impact of Mediterranean culture by the barrier of the Alps. This barrier was an advantage as well as a handicap, since it reduced the power of tradition, and thus left greater scope for freedom of thought.

As regards factual morphology, Cesalpino advanced beyond Albertus at some points, though, on the whole, he shows no great originality of observation. Elaborating an Aristotelian idea, he emphasises shoot development as a feature peculiar to plants, and contrasts it with the growth of animals, in which the parts are determined and limited from the beginning. For the process of development of the plant shoot (*germen*), he uses the word *germinatio*; this is apt to be confusing to the modern reader, since the words 'germ' and 'germination' have been invested with special meanings in the botanical terminology of to-day. Like Theophrastus, Cesalpino thinks that the foliage exists to protect flower and fruit. The belief of the Aristotelian school, that the function of leaves was merely to shelter the more delicate parts from the sun's rays, was a natural reaction to the southern brilliance of the Mediterranean climate. Cesalpino must have looked at leaves carefully, for he observed the production of buds in their axils, and he was aware that a tangled plexus of bundles may be found in the nodes. Among minor points of vegetative morphology, which he notices, are the quadrate stem of horehound (*Marrubium*), with the buds arranged, like the leaves, on four orthostichies. He discusses tendril-bearing climbers, point-

[1] Bayle, P. (1820); Renan, E. (1852), p. 332.

ing out that the leaf-stalks serve as tendrils in clematis, while ivy climbs by means of 'claws', like centipedes' feet, distributed along the whole stem. From such climbers he distinguishes the twiners, that twist round a support in the manner of a serpent; he thinks that these twiners show a certain perception of adjacent bodies, since they creep until they find, and, having found, take hold (quasi sensus quidam adiacentis corporis illis videatur inesse, cum repant, donec inueniant, et inuentum apprehendant).

Following the lead of Theophrastus, Cesalpino recognises the existence of 'leaves' with stem-like characters. It was in the genus *Ruscus* (butchers' broom) that Theophrastus had the opportunity of examining these structures in the Mediterranean region, while Cesalpino was able to add another example—*Opuntia*—from the New World flora, with which European botanists were in his time beginning to be acquainted.

About flowers, fruits, and seeds, Cesalpino recorded a number of interesting details, of which only a few can be noticed here. He studied the relation of the 'flower' (i.e. perianth, stamens and styles) to the 'seed'; the almond (*Prunus Amygdalus* Stokes) has, he says, one seed to the flower, but spurge (*Euphorbia*) has three; and, while the rose has one seed-receptacle to a single flower, bulbous plants have them in threes. In dogs'-mercury (*Mercurialis*), nettle (*Urtica*), and hemp (*Cannabis*) he recognised the existence of sterile as well as fertile flowers, and says that these sterile flowers are called male.

As regards the psyche in plants, Cesalpino's basic ideas are purely Aristotelian.[1] He works out such views, however, in considerable detail, emphasising that the form of the animal, as contrasted with that of the plant, depends to a great extent upon the possession of a sentient psyche, which is responsible for feeling and movement, and for the consequences of these two factors. He takes a special interest in the seat of the nutrient psyche in plants, and, after much discussion, adheres to Aristotle's idea that, though the psyche may in some senses be held to be distributed throughout all parts, yet the spot where root and shoot join is the most appropriate place for the psyche's special home (qua scilicet radix germini coniungitur, locus videatur cordi plantarum opportunissimus). For this *cor plantarum*—the inmost

[1] See pp. 9, 10.

citadel of plant life—he also uses the terms, root-crown (*caput radicis*) and shoot-initial (*germinis principium*). This intermediate no-man's-land between root and shoot has been called by later writers the collar, or life-knot[1] (nœud vital), and, in the seedling, the collet. De Candolle objected, in his *Organographie*, that this zone was merely "la point de démarcation de deux organes",[2] and Sachs reiterated that "this part of the plant...is really no part at all".[3] Such negative statements, however, merely indicate that nineteenth-century writers had reacted to an undue extent from the Aristotelian standpoint; for modern work on the anatomy of seedlings[4] has shown that there is a genuinely transitional hypocotyledonary region, which may be of considerable length, and which cannot be allotted consistently either to root or shoot. In other words there is, in actual fact, a boundary *segment* of interpenetration, rather than a sharply defined boundary *plane*, between the two primary regions of the plant body. Cotyledons, moreover, carry certain indications of a nature intermediate between root and shoot.[5]

Cesalpino thought that he could observe, at the junction of shoot and root, a soft substance comparable with the animal brain, which he believed to be continuous through the plant, as the spinal cord through the animal body—a view akin to that of Albertus. This notion of the pre-eminence of the innermost region of shoot and root, exercised a dominant influence upon Cesalpino's botanical theories. He followed Theophrastus in holding that seed production is the ultimate goal of the plant (in ea propagatione, quae fit ex semine, plantarum finis consistat). He linked this view with his conviction about the central seat of the vital principle, and, no doubt under the influence of Plato's notion that the sperm is derived from the marrow,[6] he concluded that the material for the seed must emerge from the pith (seminis materia...ex profundis partibus erumpit). He accounts, ingeniously, on the same ground for the fact that plants in general do not fruit from their thicker branches, but from young shoots, since the formative stuff for the seed is better able to make its way

[1] Rennie, J. (1849), p. 7.   [2] Candolle, A. P. de (1827), vol. I, p. 147.
[3] Sachs, J. von (1890), p. 47.
[4] Cf. for instance, Popham, R. A. (1947), and on budding as a common character of root and hypocotyl, Rauh, W. (1937).
[5] Cf. Arber, A. (1925), p. 177.   [6] See p. 26.

out to the surface of young shoots than through solid wood. The classical belief in the fundamental importance of the pith persisted long after the days of Cesalpino. Two centuries later, we find Hedwig inveighing against it as a widespread current error. He maintains that "the extolled virtue of the pith of plants has been attributed to this tissue without any justification".[1]

Cesalpino's idea that the vital parts of the plant are internal, leads him to the corollary that the external layers are of an inferior nature, and may even be classed as waste products; in this way he explains the fact that the surface of seeds is often black, instead of being pale like the 'pure' seed contents. That in seed production, to his thinking, the life-history came to its full beauty and perfection, vivified his interest in the actual seed itself. He re-states the Theophrastean comparison between the animal egg—consisting of the 'delineation' of the creature as it will be, with food material in addition—and the seed of a plant. He notices that the grain of wheat contains an 'eye' [embryo], which is the essential part of the seed, for, if it be destroyed, germination fails. He mentions that ants are wont to bite away this little component from the grain, before they stow their harvest underground.

It is for his belief in the value of seeds and fruits in supplying criteria for classification that Cesalpino is most frequently remembered. This principle, and that of the prime importance of the division into trees and herbs, to which he also adhered, were direct deductions from the views of the Aristotelian school. In considering classification in general, Cesalpino enumerates and rejects a number of other suggestions as to clues that might be used. Some of his rejections were justified; he recognised, for instance, the artificiality of grouping together 'coronary' herbs, merely because their colour and scent make them suitable for garlands; or such plants as turnip, aristolochia, and cyclamen, because they have a 'round root' in common. In certain other rejections he was less happy. When, for example, he ruled out leaf characters, it was because his comparative acquaintance with foliage forms was insufficient to enable him to realise their taxonomic significance, and he was thus led to the false conclusion that no such significance existed. His general tendency seems to have been to narrow the field of systematic criteria, and this was a

[1] Hedwig, J. (1781), p. 300.

31

retrograde step. The empirical development of taxonomics at the hands of the renaissance herbalists[1] gave surer guidance than Cesalpino's *a priori* views. The heart of the matter was reached only in a gradual fashion, and was epitomised by Nehemiah Grew, when he wrote, a hundred years after Cesalpino, that "the most *Philosophick* way of distinguishing or sorting of *Plants*, were by the *Characteristick Properties* in all *Parts*".[2]

[1] Cf. Arber, A. (1938a), chap. vi, pp. 163–84.
[2] Grew, N. (1682), p. 174.

CHAPTER IV

PLANT MORPHOLOGY FROM JOACHIM JUNG TO GOETHE AND DE CANDOLLE

IN 1587, four years after the publication of Cesalpino's *De plantis*, Joachim Jung was born at Lübeck. He lived until 1657, dying in the year in which Nehemiah Grew was sixteen, and Marcello Malpighi, twenty-nine; his work thus belongs essentially to the first half of the seventeenth century. His two botanical treatises, *Isagoge Phytoscopica*, and *De Plantis Doxoscopiae*, did not see the light until after his death. It appears that he left no autograph manuscript of his work on plants, but three transcripts are known, with annotations which seem to be his own. After his lifetime, several successive editors dealt with the script; the account here given is derived from S. A. Albrecht's version of 1747. Jung's closely knit, methodical style, does not lend itself to summarising; all that can be done in a brief space is to draw attention to a few of the more salient points.

Jung's general view of the plant world is based wholly upon that of the classical school. He follows Aristotle[1] and Cesalpino in the special importance which he attributes to the root-crown, in which the root and shoot meet, and for which he uses the term *fundus*. A plant with one crown, Jung describes as simple, while one with several crowns is multiplex. He considers various types of branching—that, for instance, in which one crown gives rise to one stem (unicaul, e.g. trees), and that in which it gives rise to several stems (multicaul, e.g. *Artemisia*). He notes that a unicaul plant may simulate a multicaul plant, if branching occurs not far from the crown. He points out that sometimes two kinds of stem arise from the same crown; one kind is vegetative, while the other is naked and reproductive. His account of the different kinds of shoot branching is not always accurate; but we cannot be surprised at this, since the analysis of form, as determined by

---

[1] Hett, W. S. (1935), pp. 414–15 [*On Youth and Old Age*, III, 468 b].

growth-history and branching, is even to-day an infant science in which our knowledge is markedly incomplete.

Though Theophrastus realised the differences between the upper and under surfaces of leaves, Jung surpassed him in definitely expressing the contrast between the stem, as a radially symmetrical organ, and the leaf, as dorsiventral. No such brief and simple way as this of expressing the distinction had then been devised, but Jung succeeded in making the matter clear by saying that the stem is extended in length, and shows no difference between front and back, or between right and left sides, while the leaf is extended in length and breadth from its attachment, so that the bounding surfaces of the third dimension (i.e. of the depth), differ from one another.

Leaf form was treated by Jung more systematically and in greater detail than by Theophrastus. Jung notes, for instance, that simple leaves may have the margin entire or cut, and that the cut margin may be laciniate, serrate, crenate, or dentate. He may be said to have originated much of our present-day descriptive terminology, both in connexion with leaves, and with other parts of the plant. His outlook was not, however, quite so modern as one might suppose from a glance at the list of terms used by him, which have been incorporated into botanical language, for it is often found that their meaning has since been considerably modified; he employs 'petiole', for instance, to include what is now called a 'pedicel'. As might have been expected at his date, he excelled Theophrastus in exactness and fulness of observation; whereas, for example, the Greek author merely mentions the hollowness of the onion leaf, Jung notes, in addition, the flattening of the leaf-base. In considering the general form of the laminae of leaves, he defines the *circumscriptio* as the outline from which the toothing is omitted; the *circumscriptio* of *Geranium columbinum* L. (long-stalked crane's-bill), for instance, is regarded as circular. He points out that such an outline is a mental abstraction. Jung was a philosopher, as well as a botanist, and it thus came naturally to him to realise the part played by subjective processes, even in observation which is, ostensibly, of the most objective kind.

Compound leaves aroused Jung's interest, and he distinguishes palmate (digitate, in his terminology) from pinnate forms, pointing out that both types are more common in herbs than in trees. Among pinnate leaves he notes the difference between pari- and imparipinnate, and he also draws attention to the minor components interposed between the major leaflets in *Spiraea Ulmaria, Agrimonia,* etc. These small subsidiary leaflets had been observed long before Jung,[1] but it is only in modern times that they have received full comparative study.[2] Among other details about leaves, Jung remarks upon the heterophylly of a species of *Campanula,* and of the water buttercup.

As regards leaf arrangement, Jung distinguishes stems which bear the leaves scattered (*sigillatim*), from those bearing them two, three, or more at a node (*coniugatim*). He notices that, when the stem is quadrate, all the leaves are yoked in alternating pairs; such a scheme he calls 'decussate'.

When he turns to the flower, we find that Jung defines it in a way which does not show much advance upon Aristotelian conceptions. He says that it is the more delicate part of the plant, adhering to the fruit rudiment, and distinguished by colour, or shape, or both. The flower, on his view, is thus limited to what we now call the corolla and androecium; this idea survives in popular speech to-day, where the shedding of the 'flower' means the fall of the petals and stamens. By the term 'perianth', Jung denotes that which encloses the flower—the calyx, or the involucre of bracts. This definition is more in accord with the etymology than the modern scientific usage, which treats as 'perianth', not something *around* the flower, but one of its integral components. Jung's mode of description involves his calling such a flower as that of the tulip 'naked', since it has no green calyx. As we noticed in considering his ideas about the leaf, it is remarkable how often the *words* of his terminology have survived, though sometimes changed in sense. He is speaking accurately from his own standpoint, though it sounds odd to-day, when he calls the flowers of the Polypetalae, 'composite', and those of the Sympetalae, 'simple'; for to him the flower was primarily the corolla, so it was treated as simple, if it was all in one piece, or composite, if it consisted of several free petals. He was clear, however, about the nature of those flowers which we *now* call composite, describing them as "compositus absolute ex flosculis". He not only

---

[1] See pp. 122, 123.   [2] Müllerott, M. (1940).

recognised their status, but he realised in what their 'doubling' consisted, noting that in those with disk and ray florets, doubling is brought about by encroachment of the ray florets upon the disk, obliterating the tubular disk florets.

Upon stamens Jung made some detailed observations, noticing, for instance, that the filaments, which he calls "*Pediculi Staminum*", may be glabrous, or hairy, and that they may be branched. He was interested also in the numerical relations of androecium and petals. The style he defines as the part occupying the middle of the flower, and inhering in the rudiment of the fruit or seed. He recognises that in the *Iris* the place of the style is taken by three internal erect leaves, beneath which the stamens are concealed. He emphasises the distinction drawn by the Aristotelians between the fruit and other annual parts of the plant, pointing out that leaves, flowers, and, in some cases, even shoots, fall and decay, whereas the fruit is complete in itself, and comes to perfection when it separates from the parent plant.

Every botanist who studies Jung's work must concede its solidity and usefulness, but it is possible, nevertheless, to hold that his attainments have often been over-estimated. The fact that he did not publish his botanical contributions has lent them an air of mystery which has enhanced their reputation. Jung's clear and systematic intelligence was entirely adapted for terminological description, but he does not seem to have travelled far outside this field. His morphology inclines to be static, and he lays little stress upon the cycle of change through which living things are continually passing. Owing to this limitation, the plant world assumes, under his hand, a certain aridity. In justice to him, however, we must recall that the time element does occasionally play a part in his descriptions, for example in his lucid account of the pappus of the composites, in which his meaning is best retained if we translate 'flos' as 'corolla'. He points out that "the corollas are said to be changed into pappus, but in fact no corolla is changed into pappus, but the pappus is attached to the apex of the fruit to which the corolla is likewise attached, and, as the corolla withers, dries up, and disappears, the fruit with its pappus grows, and the pappus, which previously was concealed by the corolla, is rendered more visible. Thus the pappus is produced from the corolla, as evening from afternoon."

Though this description involves a time sequence, it can hardly be said to be concerned with development, for what it sets out to demonstrate is that something which is generally regarded as a transformation, is, in reality, merely a replacement. For a more intrinsically developmental approach to the subject, we must turn to the writings of Jung's younger contemporaries, Marcello Malpighi (1628–1694) and Nehemiah Grew (1641–1712). In their day the influence of Aristotelian science was on the wane,[1] and these two workers had thus a better opportunity than their predecessors of giving free rein to their own originality. Neither Malpighi nor Grew confined his attention to the plant as it is at any given moment; they both saw it always in dynamic fashion as a flowing life-history, not as a mere cross-section of this life-history. In this respect their work recalls that of Bock (Tragus)[2] in the sixteenth century, rather than that of Jung in the seventeenth century. The developmental point of view must have been encouraged in Malpighi and Grew by their experience as physicians, accustomed to bear in mind the whole changing life-cycle of the human being, and to witness each phase passing inexorably into the next. Malpighi, indeed, did much work on the embryology of animals, so his mind was closely attuned to the developmental aspect when he turned to botany. Moreover the use of lenses, and the practice in microscopical research, which was common to Grew and Malpighi, trained and stimulated even the perception which depended upon unaided sight, so that these two workers were both versed in the subtle technique of the eye, which is needed for any appreciation of developmental processes.

In Grew's first work, *The Anatomy of Vegetables Begun*, published in 1672, and in *The Anatomy of Plants*, the final version of his writings (1682), his interest in development finds expression in a pioneer study of seedlings, especially that of the bean, while Malpighi's great book, the *Anatome Plantarum*, which appeared in two parts in 1675 and 1679,[3] and a posthumous volume (1697), include admirable and beautifully illustrated accounts of seedling structure in a large number of species. The

---

[1] No brief statement can give an adequate picture of the complex reaction from Aristotelian to Platonic and pre-Platonic views which occurred in the sixteenth and seventeenth centuries; see Johnson, F. R. (1937).

[2] Arber, A. (forthcoming).   [3] See also his collected works, 1687.

stress which he laid on the developmental aspect of this study is shown by the fact that he gives no less than twenty figures illustrating the seed and seedling of the castor-oil plant (*Ricinus communis* L.) at different stages, while three plates are devoted to a large number of drawings of the different phases lived through by the seedling of the date palm (*Phoenix dactylifera* L.). In wheat (*Triticum* sp.) he describes the embryo and the seedling in several stages during the first four days, and then at about six days, eleven days, and a month, thus recognising the full significance of the time element in such sequences.

Both Malpighi and Grew studied not only the development of the plant as a whole, but also that of the individual shoot. Grew, for instance, showed that he understood the nature of buds, when he noted that the "two extraordinary [additional] small Plumes" in the axils of the cotyledons of the "great Garden-Bean", differ from the terminal bud of the main shoot "in nothing save in their size".[1] This means that he had recognised the essential identity of shoots, whether primary or lateral. Malpighi, also, gave serious consideration to shoot buds,[2] and realised that they were in essence individual plants. He describes them metaphorically as infants, which will eventually 'adolesce' into branches (*Gemmae* igitur sunt velut infans custoditus, qui tandem adolescit in ramum); he compares them, moreover, to embryos produced from eggs. He realises that structurally they represent the plant "in little, not yet, as it were, unfurled" (compendium sit plantulae nondum explicitae). He calls the attachment of the bud 'implantatio', which indicates that he visualised the bud itself as an independent entity. In considering the individual leaf, Grew, by a happy inspiration, grasped the fact that vernation was actually part of ontogeny; this idea was lost to sight after his time, and was only rediscovered in the nineteenth century.[3] In Grew's words, "The Formations and Fouldings of Leaves have one Date, or are the contemporary works of Nature; each Leaf obtaining its distinct shape and proper posture together."[4] In his studies of the earlier stages of leaf development, Malpighi surpasses Grew. In the elm (*Ulmus*), for instance, his figures show the relatively large size of the stipules in extreme youth,

[1] Grew, N. (1672), p. 9.   [2] Malpighi, M. (1675), *Idea*, p. 6.
[3] Cf. Arber, A. (1941 b), p. 236.   [4] Grew, N. (1672), p. 114.

and he also recognised the lateness of petiole development.[1] Moreover he seems to have observed the marginal type of lamina-production, which is now called pleuroplastic. His interest was not confined to the stages passed through by the leaf between the first rudiment and maturity; he also studied the *seriation* of leaves in the shoot. In the bud of the almond (*Prunus Amygdalus* Stokes), amongst other plants, he figures the graded leaf series from the bud-scale to the mature form of foliage-leaf, showing the gradual appearance of the petiole, as well as of the lamina, which, in the outer leaves is a mere point.[2] He thus recognised a set of intergrades between fully developed foliage-leaves, and scales which carry little suggestion of typical leaf characters. Nehemiah Grew pursued a corresponding train of thought, which led him to identify, as Malpighi[3] also did, certain members still more remote from one another at first glance—namely foliage-leaves and floral leaves. This conception, indeed, is found in classical writings. Theophrastus employed the word leaf (φύλλον) for a petal—a use which we repeat to-day when we talk about 'a crumpled rose leaf'. Moreover it must have continued through the medieval period, if we may judge by Dante's expression, 'flowers and other foliage'.[4] Such instinctive comparisons of perianth members with leaves were developed by Grew on reasoned lines. He points out that the sepal, or 'Empaler', as he calls it, has the same anatomical construction as a foliage-leaf, since it consists of "the *Skin*, the *Cortical* and *Lignous Bodies*; each *Empaler* (when there are divers) being as another little Leaf".[5] He notes that those of the quince blossom (*Cydonia oblonga* Mill.) "thrive so far as to become handsom Leaves, continuing also after the *Flower* is fallen, firm and verdent a great while".[6] He takes a corresponding view of the corolla, which he calls the 'Foliation'; he writes that it is "of the same substantial nature with the Green Leaf; the *Membrane*, *Pulp*, and *Fibres* whereof, being, as [in the Empalement] but the continuation of the *Skin*, the *Cortical* and *Lignous Bodies*".[7] We see from

[1] Malpighi, M. (1675), pl. X, fig. 52.   [2] Malpighi, M. (1675), pl. XI, fig. 55.
[3] Malpighi, M. (1675); e.g. p. 42.
[4] *Purgatorio*. Canto xxxii, 38–9.   "una pianta dispogliata
                                di fiori e d'altra fronda"
[5] Grew, N. (1672), p. 129.   [6] Grew, N. (1672), p. 139.
[7] Grew, N. (1672), p. 132.

these remarks that Grew regarded not only external form, but also anatomical structure, as morphological features. He was thus in agreement with Adrian Spieghel, who had written, more than sixty years earlier, that by the 'form' of a plant he meant the shape or make, as much internal as external, of the body in which the vegetative psyche inheres (sed figuram seu faciem tàm internam, quàm externam corporis, cui haec anima vegetativa inhaeret).[1]

A century after Grew had demonstrated the foliar nature of sepals and petals, C. F. Wolff carried the matter to a further point by maintaining that not only these members, but also stamens and pericarp segments, were modified leaves.[2] This seems to have been the first definite statement that *all* the parts of the flower are to be regarded as leaves. That this thesis is at least a step towards the truth, is suggested by the fact that it has been reached independently by so many botanical thinkers. The poet Goethe,[3] for instance, in a little book—*Versuch die Metamorphose der Pflanzen zu erklären*—published in 1790, more than twenty years after Wolff's paper, put forward ideas closely related to those of Grew and of Wolff, though he was unacquainted at that time with the work of either of them. Goethe's ignorance of his predecessors' writings was probably an advantage, since his belief, that his theory was wholly his own, led him to develop it with special verve and freshness, and to give it a certain unity of presentation which it might otherwise have missed. His tendency to disregard the relevant literature was one of the results of his amateur approach, which also showed itself in incompleteness of factual knowledge. In matters of detail he was never, indeed, fully abreast of his time. This may be the reason why his thought—recalling, in this respect, that of Francis Bacon—had so curiously little effect on the science of his own period; but, like Bacon's, it has been a source of lively stimulus to succeeding generations, to whom his

---

[1] Spieghel, A. (1606), cap. II, p. 4.

[2] Wolff, C. F. (1768); see pp. 404–6.

[3] For a fuller study of Goethe's contribution to morphology, with a translation of the *Metamorphose*, see Arber, A. (1946a); cf. also pp. 59 *et seq.* of the present book. After the *Metamorphose*, Goethe produced no other complete botanical treatise, but he left essays and fragmentary notes, which are collected conveniently, and annotated, in Troll, W. (1926); they will also be found in more complete form in the definitive Sophien-Ausgabe (Weimar) of Goethe's works; Goethe, J. W. von (1887 etc.).

mistakes in matters of fact have ceased to be of the slightest consequence, and who recognise his gift of insight, as de Candolle did when he wrote that "M. Goethe... a comme deviné l'organisation végétale".[1] Goethe himself realised that the value of his work did not lie in encyclopaedic acquaintance with the subjects he discussed, but rather in the unifying quality of his mind. As he said, "To pursue botany further into details is not in my line. I leave that to others who far surpass me therein. My only concern was to trace back the separate phenomena to a general and fundamental law."[2] It was an inevitable result of this attitude that he tended to prune the issues involved, until a point of over-simplification was reached. For instance, in his *Metamorphose*, he practically disregarded every form of flowering plant except the annual, and he scarcely considered the root at all. Moreover, as Turpin[3] pointed out, the title of Goethe's essay laid claim to a field which was not, in fact, covered, for it should have been described as *Metamorphosis of the appendicular organs of the shoot*, rather than as *Metamorphosis of plants*. Goethe did, however, to some extent consider the shoot in general; he followed Malpighi and earlier writers in recognising that the lateral branches springing from the nodes may be regarded as individual plantlets, which bear the same relation to the mother-plant, as the mother-plant to the earth in which it is rooted.

We do not depend upon mere conjecture about Goethe's approach to phytology, since he wrote his own botanical history in autobiographical form.[4] He tells us that he plunged fervently into Linnaean botany, and that, at the time of his greatest enthusiasm, he carried certain works on the subject with him everywhere. Even at this period, however, the Linnaean attitude to plant science failed to give him entire satisfaction, since he was greatly troubled by the versatility of plant members, such as the leaves borne on a single stem, which might present a series of different forms, grading into scales, and finally disappearing altogether.

[1] Candolle, A. P. de (1827), vol. I, p. 243.
[2] Eckermann, J. P. (1836–48), vol. I, p. 337, 1 Feb. 1827.
[3] Turpin, P. J. F. (1837), p. 7.
[4] Goethe, J. W. von (1831), pp. 107–63. *Der Verfasser theilt die Geschichte seiner botanischen Studien mit* (German and French versions). German version also in Troll, W. (1926), pp. 187–209.

It was under the stirring influence of a journey across the Alps into Italy, that the theory, which he felt be the solution of all such difficulties, presented itself to his mind. The richness of an exotic flora (die Fülle einer fremden Vegetation) was first brought home to him in the botanical garden at Padua—the oldest in Europe. Here he happened to see a palm (*Chamaerops humilis* L.) offering all gradations from the first lanceolate leaves up to the complex mature form, and then the sudden transition to the spathe and inflorescence. The gardener gave him a suite of specimens from this tree, which he carried away in triumph. Some thirty years later, he wrote that he still had this set of leaves before him, and that he still looked upon them with veneration because of the part that they had played in the development of his ideas. The actual palm tree in question is said to date from 1584, and in the twentieth century it still flourishes.[1] It was in contemplating such phenomena as the leaf series in *Chamaerops* that Goethe was led to the notion that all the flowering plants were comparable, and that they could be brought together under a single concept (sie sich nun unter einen Begriff sammeln lassen). Then at last, after, in the spring of 1786, he had reached Sicily—the ultimate goal of his excursion—the original identity of all the lateral appendages belonging to the shoot became completely clear to him (leuchtete mir... die *ursprüngliche Identität* aller Pflanzentheile vollkommen ein); as he wrote subsequently, "When now the plant vegetates, blooms, or fructifies, so it is still *the same organs* which, with different destinies, and under protean shapes, fulfil the part prescribed by Nature."[2] For the type form which manifests itself in all these organs, he used the word 'leaf' (Blatt)—a use which is obviously open to serious objection, since it immediately brings a foliage-leaf to mind. This objection was felt to some degree by Goethe himself, but, if he had been aware of its full force, he would no doubt have found an alternative term of a more generalised character, such as the word 'phyllome', which became current in the nineteenth century. Though, as we have seen, Wolff had to a considerable extent anticipated Goethe's views, Wolff's version is relatively crude, in so far as he interprets the parts of the flower naïvely as modified *foliage-leaves*. Goethe had

[1] Information by letter, dated 14 Sept. 1945, from Prof. G. Gola, Istituto Orto Botanico, R. Università di Padova.   [2] Goethe, J. W. von (1790), § 115.

advanced much further, in the realisation that the foliage-leaf had no claim to be regarded as the type appendage, since the type was a mental concept, not to be imprisoned within the limits of any one kind of actual member.[1]

When we follow Goethe's account of the sequence of stages in plant growth, we find that he detected the foliar character of cotyledons, even when this was by no means obvious. "They often appear", he says, "shapeless, crammed, as it were, with crude matter, and as much extended in thickness as in breadth.... These cotyledons bear scarcely any resemblance to a leaf, and we may be misled into taking them for organs belonging to some special category (besondere Organe)"; but he adds that "Nevertheless in many plants they approach leaf form; they increase in area and become thinner; when exposed to light and air they assume a deeper green; and the vessels [bundles] which they contain become more recognisable, and more similar to the veins of a leaf." And he concludes, "Finally they appear before us as true leaves.... Their resemblance to the succeeding leaves prevents our taking them for special organs; we recognise them, rather, as the first leaves of the stem." He proceeds to show that the plane of their attachment must be treated as the first node. Passing upwards from the cotyledons, Goethe observed that the earlier members of the plumule are generally typical foliage leaves, but that, like the cotyledons, their form tends to be simpler than that of the mature leaves. The succeeding leaves show more and more elaboration, and, soon after they have reached their greatest expansion and development, "we become", he writes, "aware of a new aspect which warns us that the epoch which hitherto we have been studying is over, and a second is approaching—the epoch of the *Flower*.... We see the transition to anthesis come to pass either *relatively rapidly* or *relatively gradually*. In the latter case we commonly notice that the stem-leaves begin to draw in, as it were, from the periphery, and especially to lose their diverse marginal divisions, while, on the other hand, they show some expansion in their basal regions, where they are connected with the stem." Where the transition to the reproductive region is abrupt, "the stem, above the node of the uppermost leaf, suddenly becomes tall and slender".

[1] Cf. also pp. 59 *et seq.*

The key to Goethe's position regarding the relation of floral and foliar members is given by his statement that "It may, it seems to us, be proved most clearly that the leaves of the calyx[1] are just the same organs as those which up to the present have developed as stem-leaves, but are now, often in very different guise, collected round a common centre." Goethe comments on the whorling of the sepals as resembling that of the cotyledons in conifers. This whorling of leaf structures may, he notes, often be associated with "concrescence of their lateral margins". He relates this to close crowding and pressure in the embryonic condition. He observes that in a bell-shaped calyx, produced by such fusion, the upper margin is more or less toothed or divided, in accordance with the composite origin of the structure.

That petals, and what we should now call perianth members, share leaf nature with the sepals, Goethe considers to be indicated by the fact that, in the neighbourhood of the inflorescence, appendages may occur, which obviously belong to the foliage-leaf category, but yet are coloured (bracts). Moreover "stem-leaves may pass into petals. So, for example, an almost completely developed and coloured petal may often be found on the stem of a tulip." If the foliar nature of petals be accepted, that of stamens follows as a matter of course; indeed, as Goethe writes, "Were the relationship of all the other parts to one another so obvious, so generally observed, and so indubitably settled, the present treatise might be held to be superfluous." He points to the petal-like stamen of *Canna*, but he bases his view chiefly upon the evidence of abnormalities; he describes in particular, the petaloid stamens of double varieties of rose and poppy. He discusses, also, certain forms of staminode and nectary, which he regards as intermediate between petals[2] and stamens.

When he turned to the gynaeceum, Goethe was faced with a much more difficult task, since he was hampered by lack of understanding of the actual construction of this organ. His detailed interpretations are thus liable to be wrong, but, at the same time, he succeeded in illuminating some of the more salient points. He

---

[1] It should be noted that, in Goethe's usage the word *Kelch* (calyx) sometimes meant an involucre of bracts, and sometimes a whorl of sepals.

[2] In Goethe, J. W. von (1790), p. 51, the word *Kelchblättern* seems to be a slip for *Kronenblättern*.

visualises the legume as "a simple folded leaf concrescent by its margins", and he says that "compound seed-vessels would be explained as consisting of several leaves united round a middle point, their inner faces open towards one another, and their margins fused". He notes the petaloid character of the stigmatic region in the flower of the *Iris*—a feature to which Jung had previously referred—and the leaf-like stigma of *Sarracenia*. He observes also that styles and ovaries may become petal-like in abnormal flowers; and he has seen in the double poppy that "the stigmas of the seed-capsule are transformed into delicate little coloured leaves, completely resembling petals".

Goethe carried his study beyond the individual members to the flower as a whole, on the nature of which he held a perfectly explicit view. He says that "flowers which develop from lateral buds are to be regarded as entire plants, which are set in the mother-plant, as the mother-plant is set in the earth". This is precisely what he says, also, about vegetative shoots; it is thus evident that he equated vegetative shoots and flowers. Moreover he recognised the essential distinction that the flower differs from the vegetative shoot in being a branch *of limited growth*. He expresses this idea by saying, "a plant which *flowers* has contracted all its parts; increase in height and breadth is, as it were, arrested; and all its organs are in a highly condensed state, and developed in close proximity to one another". In abnormal flowers he finds that this limitation may fail to operate; he refers, for instance, to a rose in which the axis was prolonged through the flower, and bore petals, and foliage-leaves, and even imperfect lateral rosebuds.

Keeping this idea of the equivalence of lateral shoot and simple flower firmly in mind, Goethe arrived at a clear picture of the construction of the composite 'flower'. "Nature", he says, "forms a *common calyx* [involucre] from *many* leaves, which she presses upon one another and collects round an axis. With the same strong growth-impetus she modifies *an elongated stem*, as it were, in such a way that *all its buds are produced at once in the guise of flowers, thronged together in the closest possible proximity*.... In this monstrous crowding, the nodal leaves do not invariably disappear; in the thistles the bract faithfully accompanies the floret, which develops from its associated bud."

Closely similar views on the flower[1] to those of Goethe were developed by its younger contemporary, Augustin-Pyramus de Candolle (1778–1841), who was born in Goethe's twenty-ninth year. De Candolle was a professional botanist, who accomplished a vast amount of solid taxonomic work, so that his knowledge of plants was incomparably more extensive than that of the poet; but de Candolle's first enthusiasm was never depressed by his load of learning, and he approached morphology with as much freshness of mind as Goethe. His life was passed in French Switzerland and in France, and he was unable to read German;[2] there is indeed no evidence that he had any contact with Goethe's ideas until in 1823 he received a French 'extrait' of the *Versuch die Metamorphose der Pflanzen zu erklären* of 1790.[3]

Like Goethe, de Candolle recognised that bracts are equivalent to leaves, and also that they approach sepals in character, and he concludes that "la transition des organes de la végétation à ceux de la floraison se trouve ainsi tellement graduée, que plus on l'étudie, plus on arrive à comprendre cette unité de composition qui fait le base de l'organographie philosophique". Like Grew, he was led to equate leaves, bracts, and sepals, partly on the ground of the resemblance of their internal anatomy. He argues further that their equivalence is confirmed by the fact that leaves and sepals decompose carbonic acid gas, and exhale oxygen, when placed under water in sunlight, and that they are both liable to etiolation in darkness. He regards petals and stamens as of the nature of leaves; "j'observe", he writes, "que dans tous les végétaux connus, ces deux organes ont la même origine, et je conclus, par une analogie très-puissante, que les pétales des plantes ne sont, en thèse générale, que des filets d'étamines, ou développés par suite de l'avortement de l'anthère, ou dont l'anthère a avorté par suite du développement du filet". The carpel he considers to be "une petite feuille courbée ou pliée en-dedans sur elle-même". Like Goethe, again, he regarded the flower as a whole, "considérée sous le rapport organographique", as consisting of "l'assemblage de plusieurs...verticilles de feuilles, diversement transformées, et situées en forme de bour-

---

[1] For these views, see Candolle, A. P. de (1813) and (1827), *passim*.
[2] Candolle, A. P. de (1827), vol. I, p. 7, footnote.
[3] Candolle, A. P. de (1862), pp. 572–3.

geon à l'extrémité d'un rameau". Elsewhere he notes that, as compared with the vegetative shoot, the flower has its axis abbreviated to a point at which the internodes become indistinguishable.

An outline sketch, so brief as that just given, cannot do justice to the work of Goethe and of de Candolle on the morphological nature of the flower. Instead, however, of dealing with their views in fuller historical and critical detail, it will be more worth while to marshal the evidence as it presents itself to us to-day, with a view to determining how far the Goethean standpoint should still be adopted.[1] It will be necessary, for the sake of our argument, to devote the rest of this chapter to that purpose, though it will involve covering ground which may strike the reader as over-familiar.

That there is, as Goethe and de Candolle thought, a common basic scheme of organisation, underlying both the leafy shoot and the reproductive system, would seem too obvious to stress, if it were not that it has sometimes, in recent years, been repudiated. It has been claimed, for instance, that leaf- and flower-buds diverge in certain ontogenetic characters, which are of prime significance.[2] Newer work, however, tends to show that the development is really comparable in both.[3] In their external features, there can be no doubt that reproductive shoots often closely resemble the vegetative shoots of the same plant in their general habit and mode of branching. For instance, the 'tillering' or 'pompon' growth of certain grasses is equivalent, vegetatively, to the repetitive branching which, in the same family, gives rise to clustered inflorescences.[4] This, however, is a resemblance of a highly generalised kind, and we must now consider whether evidence of a detailed sort can be brought forward, which may make the theory of the actual equivalence of the flower and the vegetative branch acceptable to-day. On comparing the lateral vegetative shoot and the flower (using this term in a wide sense, to include pedicel and bracteoles), we find that they agree in

[1] The evidence for Goethe's theory is more fully developed in Arber, A. (1937a); see pp. 162 et seq.
[2] Grégoire, V. (1938), p. 332 etc.
[3] See, for instance, Miller, H. A. and Wetmore, R. H. (1946), p. 8; Arnal, C. (1945), p. 153; Philipson, W. R. (1946), p. 267, and (1947b), pp. 190, 191.
[4] Arber, A. (1934), pp. 215, 261-2.

arising in the axil of a leaf member. This member, in the vegetative shoot, is usually a foliage-leaf, and, in the flower, a bract; but there are numerous normal and abnormal examples in which foliage-leaves and bracts grade into one another. One of these is illustrated in Fig. 1, A 1—A 4, p. 49. A 1 is a normal head of oxeye daisy, which was borne on a long stalk, while A 3 is a sterile head, sessile and vegetative, from the same shoot. The normal inflorescence has an involucre of bracts with scarious margins (A 2), while, in the abnormal inflorescence, transitions are to be traced between the foliage-leaves and bracts recalling those of the involucre (A 4).

When we pass from the axillant phyllome to the branch it subtends, we find that vegetative shoot and flower each consist of an axis of similar structure bearing single or paired first leaves, which, though they happen to be distinguished as prophylls in the vegetative shoot and bracteoles in the flower, show an evident correspondence. In the monocotyledons, where one adaxial prophyll is the rule, we find one adaxial bracteole, while in the dicotyledons, whose paired prophylls are placed to right and left, there are equivalent paired bracteoles, one on either side. It seems then, that, in the architecture of the system of which it forms a part, and in its first leaves, the flower obviously resembles the vegetative shoot; but it is when the region above the prophylls or bracteoles is considered, that the differences between the vegetative shoot and the flower become more evident. The primary divergence is, as Goethe emphasised, that the axis of the vegetative shoot tends to elongation, so that the leaves are separated by well-defined internodes, whereas the floral axis tends to be telescoped from the level of the lowest sepal. In 1759, long before Goethe's work on metamorphosis, Wolff had written that the calyx consists of the last leaves which are put forth before the inhibition (Verstopfung) of the growing point.[1] It is true that there are occasional exceptions to the abbreviation of the floral axis, exceptions which bridge the gap between floral and vegetative shoots. As well-known examples we may cite *Gynandropsis speciosa* DC., in which there are prolonged internodes between corolla and androecium, and between androecium and gynaeceum (Fig. 1, D, p. 49), and also the less striking but more

[1] Samassa, P. (1896), Th. I, § 121, p. 69.

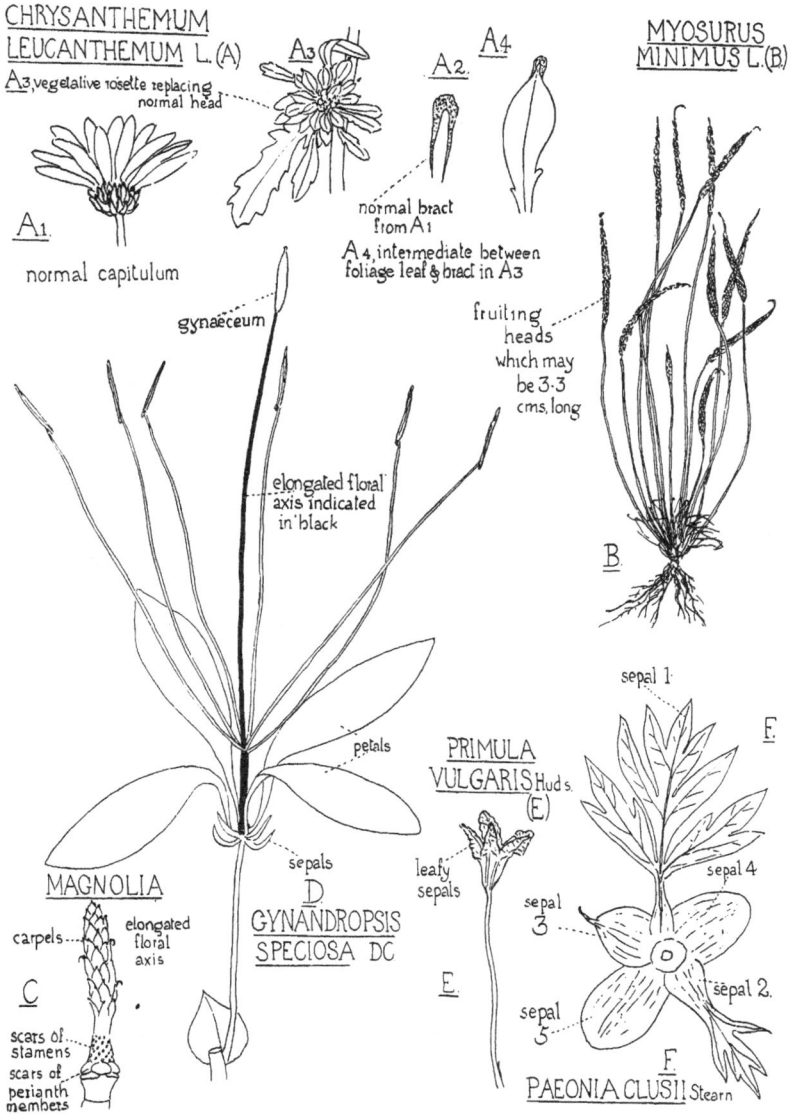

CHRYSANTHEMUM
LEUCANTHEMUM L. (A)

A3, vegetative rosette replacing
normal head

A3

A2.

A4

A1.

normal capitulum

normal bract
from A1

A4, intermediate between
foliage leaf & bract in A3

MYOSURUS
MINIMUS L. (B)

gynaeceum

fruiting
heads
which may
be 3.3
cms. long

elongated floral
axis indicated
in black

B.

petals

PRIMULA
VULGARIS Huds.
(E)

sepal 1

F.

MAGNOLIA

carpels

elongated
floral
axis

sepals

D
GYNANDROPSIS
SPECIOSA DC

leafy
sepals

sepal
3

sepal 4

C

scars of
stamens
scars of
perianth
members

E.

sepal
5

sepal 2.

F.
PAEONIA CLUSII Stearn

Fig. 1. Vegetative features in floral shoots. A 1, A 3, B, C, E, F ($\times\frac{1}{2}$):
A 2, A 4 (enlarged); D (nat. size).

familiar instances of the elongated receptacle in magnolia (Fig. 1, C) and mousetail (*Myosurus minimus* L., Fig. 1, B). These exceptions to the contraction of the floral axis are, however, so rare that the best term of comparison for the flower is, not a mature vegetative shoot, but a vegetative *bud*; the flower might, indeed, be described as corresponding to a vegetative shoot in a condition of permanent infantilism. It has to be borne in mind that the comparison is not between the flower and a vegetative shoot in which the apical activity has been exhausted, but between the flower and a still active shoot, whose activity has, however, ceased to express itself in apical extension, but has become diverted into other channels. The frequent elongation of the internode below the calyx is perhaps one of the correlatives of the cessation of apical growth in the floral axis. The fact that the normal order of development of lateral appendages (older below, younger above) is sometimes disturbed in the flower by the precocious appearance of those nearer the apex,[1] may be regarded as coming into the same category. A third feature, which may also be related to checked apical growth, is the development of perigyny and epigyny.

When we turn from the general make-up of the floral shoot to its component parts, we find that there is little difficulty in equating the bract and sepal with foliage-leaves, as Goethe suggested; their resemblances are so conspicuous that it seems quite natural to class them together, especially as perfect series can be traced connecting these three members. A thorough comparative study by Glück led him to the conclusion that a sepal may correspond either to the whole of a stipulate foliage-leaf, or to certain elements of it;[2] so we need not do more here than refer to a few illustrative examples. In *Paeonia Clusii* Stearn (Fig. 1, F, p. 49) the outer sepals may have an elaborate foliaceous blade. The sepals of the rose, also, grade from branched structures, recalling foliage-leaves, into simple, typically sepaline forms. This peculiarity has long been noticed: Sir Thomas Browne[3] described the "strange disposure of the Appendices or Beards" in "the five Brethren of the Rose", as they are called in an old rhyme. Fully foliar

[1] Goebel, K. von (1933), Ed. 3, pt. III, p. 1837.
[2] Glück, H. (1919), p. 363.
[3] Browne, Sir T. (1658), *The Garden of Cyrus*, p. 140.

sepals occur moreover as an abnormality in many other genera, such as *Geum* (Fig. 4, A, p. 54), *Primula* (Fig. 1, E, p. 49), and *Rubus* (Fig. 6, C2–C4, p. 81). We may recall, also, that sepals may, though rarely, resemble foliage-leaves in the possession of stipules, e.g. *Periploca graeca* L. (Fig. 2, B, p. 52). It is possible that the exstipulate character generally observed in floral members, is correlated with the abbreviation of the axis. This is suggested by the fact that in the hawthorn (*Crataegus Oxyacantha* L.), and other woody plants,[1] we find long shoots, in which the leaves have well-developed stipules, and also short shoots in which stipules are absent (Fig. 2, A, p. 52). As well as with foliage-leaves, sepals may reveal an affinity with petals. For instance, in *Chimonanthus fragrans* Lindl., there is no sharp distinction of calyx and corolla, but we may trace a series of transition forms between scarious bracts, and perianth members, of which the outer are broadly based, while the inner are spathulate and more petal-like (Fig. 4, B1–B3, p. 54). In certain plants, an abnormal transformation of sepals into petals may be witnessed; in Canterbury-bells, for instance, the green calyx may be represented by an outer corolla (Fig. 3, B1–B2, p. 53). Certain resemblances are even traceable between normal petals and foliage-leaves of a simple type. The petal of *Polanisia viscosa* DC. illustrated in Fig. 3, A1–A3, p. 53, shows leaf-like characters in its venation and in the possession of stomata. In general, however, the resemblance to foliage-leaves diminishes in the floral members succeeding the calyx. It may be recalled that Schoute,[2] from his prolonged study of floral whorls, concluded that the calyx is akin in its phyllotaxis to the vegetative region— for the calyx-spiral in terminal flowers continues the phyllotaxis of the foliage-leaves, and, in lateral flowers, that of the prophylls —while it is the corolla in which typical whorl-formation first appears. It is thus above the sepals that the change from the vegetative to the reproductive phase declares itself, in the scheme of arrangement of the appendages. The corolla, moreover, in its general characteristics, is related to the androecium that succeeds it, rather than to the calyx that precedes it. It is not necessary to labour this point, since the evidence is fully familiar; it is illus-

[1] Troll, W. (1935, etc.), I. 2. 2, p. 1276.
[2] Schoute, J. C. (1935), Introd. and p. 4.

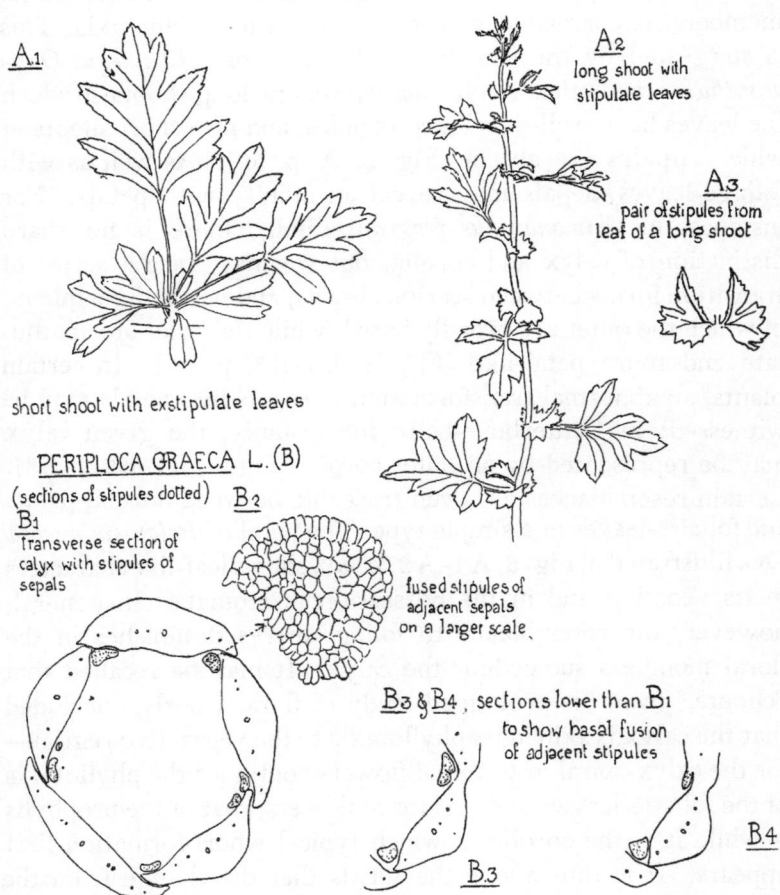

CRATAEGUS OXYACANTHA L. (A)

A1

A2
long shoot with
stipulate leaves

A3.
pair of stipules from
leaf of a long shoot

short shoot with exstipulate leaves

PERIPLOCA GRAECA L. (B)
(sections of stipules dotted)
B1
Transverse section of
calyx with stipules of
sepals

B2

fused stipules of
adjacent sepals
on a larger scale

B3 & B4, sections lower than B1
to show basal fusion
of adjacent stipules

B3

B4

Fig. 2. Parallelism between flower and vegetative shoot. A 1–A 3 (×½);
B 1, B 3, B 4 (× 23); B 2 (× 193 *circa*).

52

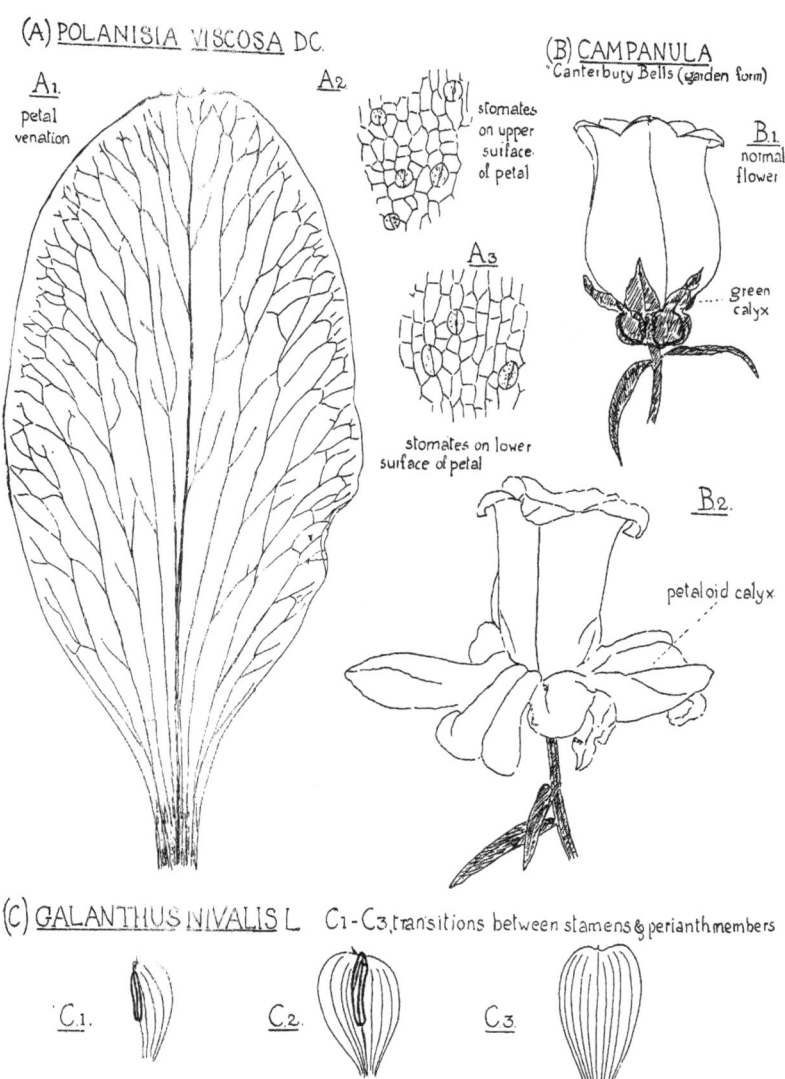

(A) POLANISIA VISCOSA DC.

A1.
petal venation

A2.
stomates on upper surface of petal

A3.
stomates on lower surface of petal

(B) CAMPANULA
Canterbury Bells (garden form)

B.1
normal flower

green calyx

B2.
petaloid calyx

(C) GALANTHUS NIVALIS L. C1–C3, transitions between stamens & perianth members

C.1.   C.2.   C3.

Fig. 3. Transition between floral and vegetative characters in the flower.
A 1 (× 14); A 2, A 3 (× 193 *circa*); B 1, B 2 (× ½); C 1–C 3 (enlarged).

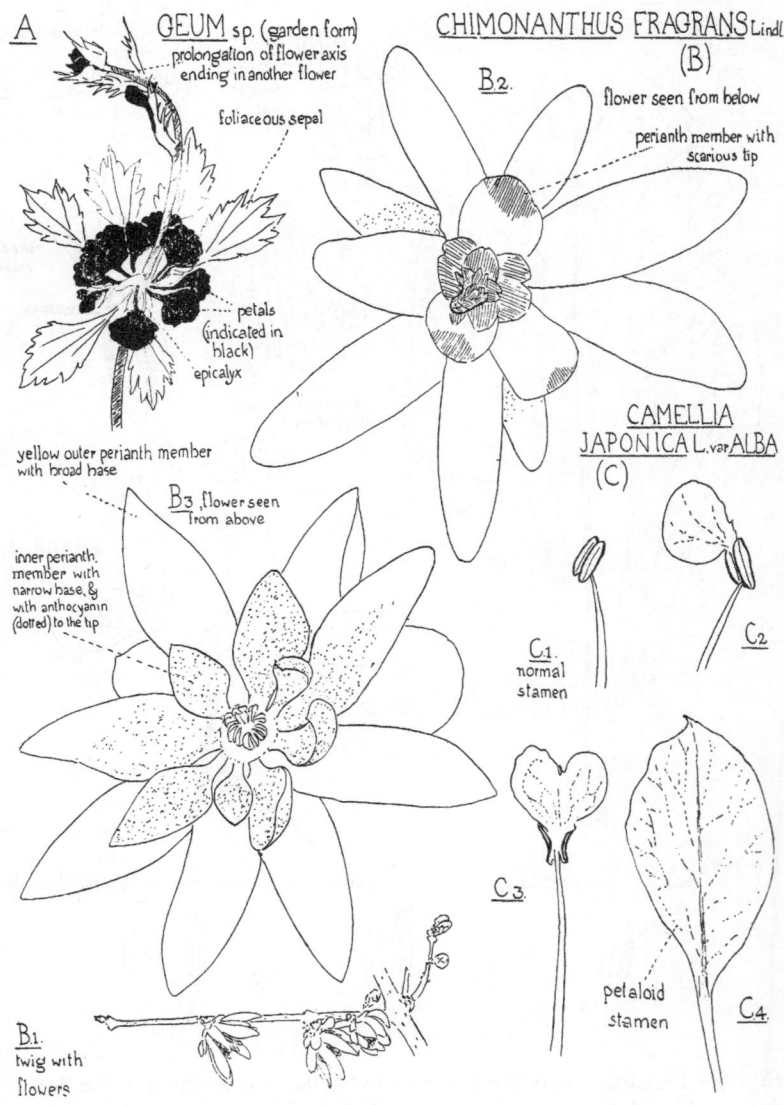

A

GEUM sp. (garden form)
prolongation of flower axis
ending in another flower

foliaceous sepal

petals
(indicated in
black)
epicalyx

CHIMONANTHUS FRAGRANS Lindl.
(B)

B2.

flower seen from below

perianth member with
scarious tip

CAMELLIA
JAPONICA L. var ALBA
(C)

yellow outer perianth member
with broad base

B3, flower seen
from above

inner perianth
member with
narrow base, &
with anthocyanin
(dotted) to the tip

C1.
normal
stamen

C2

C3

petaloid
stamen

C4.

B1.
twig with
flowers

Fig. 4. Transitions between vegetative and reproductive members in the
flower. A, B 1 (×$\frac{1}{2}$); B 2, B 3, C 1–C 4 (enlarged).

trated by certain abnormal stamens showing petalody of the connective in camellia (Fig. 4, C, p. 54) and snowdrop (Fig. 3, C, p. 53). We may indeed agree with Goethe and de Candolle that petals and stamens show so much affinity that it is evidently reasonable to group them together, and to look upon the corolla-androecium system as jointly forming the male part of the flower. The petals will then be regarded as transition members between the vegetative and the actively reproductive parts of the floral shoot. Possibly the word *transition* is out of place, and we are dealing, actually, with a *mixture* of characters in varying proportions. Malpighi,[1] who, long ago, drew attention to intermediates between petals and stamens in the rose, regarded their structure as "mixtura staminis et folii"; that is to say, he looked upon them as representing an intermingling of two distinct natures, rather than as phases in a process of metamorphosis. There is indirect evidence for such a possibility in modern work,[2] which has demonstrated that, in rye plants (*Secale cereale* L.), in which the earlier lateral rudiments from the shoot apex produce leaves, and the ultimate rudiments, spikelets, there is an intermediate region in which the type of outgrowth is dependent upon conditions. This is not, however, as might be supposed, because the rudiments in this region are genuinely transitional; they seem, on the contrary, to be of duplex structure, and it has been suggested that they consist of a spikelet initial associated with a leaf initial. Conditions thus appear to effect no transformation—they merely determine which kind of initial shall develop, and which shall remain in abeyance. If we interpret petals in the light of this, admittedly remote, analogy, we should say that, in them, both the male character and the vegetative character are present in a partnership, in which vegetativeness is predominant, while, in the stamens themselves, the vegetative character normally remains latent, while the male character reaches full expression.

The direct comparison of the androecium with foliage-leaves is rendered less easy by the extreme simplicity of the stamen vascular system, which usually takes the form of a single strand. Occasionally, however, the anatomy is more fully developed. In

[1] Malpighi, M. (1675), p. 46 and pl. 28, fig. 160.
[2] Purvis, O. N. and Gregory, F. G. (1937), pp. 583 and 590.

Bauhinia yunnanensis Franch. (Fig. 5, A 1–A 3, p. 57), for instance, and in *Amherstia nobilis* Wall., at the attachment of the anther two branches are given off from the filament bundle; these branches pass down into the two anther lobes. There are thus three bundles—a midrib and two laterals—corresponding to the basic scheme in many foliage-leaves. Rarely the filament itself shows a structure that may be compared with that of a petiole; the bamboo, *Ochlandra setigera* Gamble, for instance, has a median and two lateral bundles in this region.[1] Moreover there are certain examples in which there is an actual bundle-ring in the filament, as in *Amherstia nobilis* Wall. (Fig. 5, C, p. 56) and *Cassia bicapsularis* L. (Fig. 5, B 1–B 2). It is, however, uncertain whether much stress can be laid upon the petiole-like character of these examples from the Leguminosae, since in *Amherstia* there is evidently a strong tendency throughout the bracteoles and the flower to the replacement of simple strands by concentric bundles or bundle rings.[2] As an instance which is not open to this objection we may cite the caper (*Capparis spinosa* L.), in which the filament anatomy might well pass for that of a leaf-stalk (Fig. 5, D).

Then we turn to the carpels, we find that their striking variety of form, and their tendency to fuse—among themselves, or with the residual apex of the floral shoot—into a complex gynaeceum, make it difficult, at first glance, to visualise them all in any unified way. Goethe and de Candolle's view, however, which treats them as equivalent, individually, to foliage-leaves, has proved in general remarkably effective in the reduction of all this apparent heterogeneity to an orderly plan. The key to this plan, expressed in structural terms, is that the carpel may be compared with a leaf member possessing a midrib and two main lateral veins; the ovules are, as a rule, produced on the upper surface, in the neighbourhood of the lateral margins, and in association with the main lateral veins. When there are fusions of the incurved carpellary wings, either with each other, or with those of adjacent carpels, these fusions may occur along a line more or less remote from the actual edge, so that it is the external surfaces which unite. A hint of the foliar nature of the carpels, and of the marginal location of

[1] Arber, A (1929*b*), fig. 7, A 2, C 1, p. 775; or (1934), fig. 47, A 2, C 1, p. 125.
[2] Arber, A. (1940).

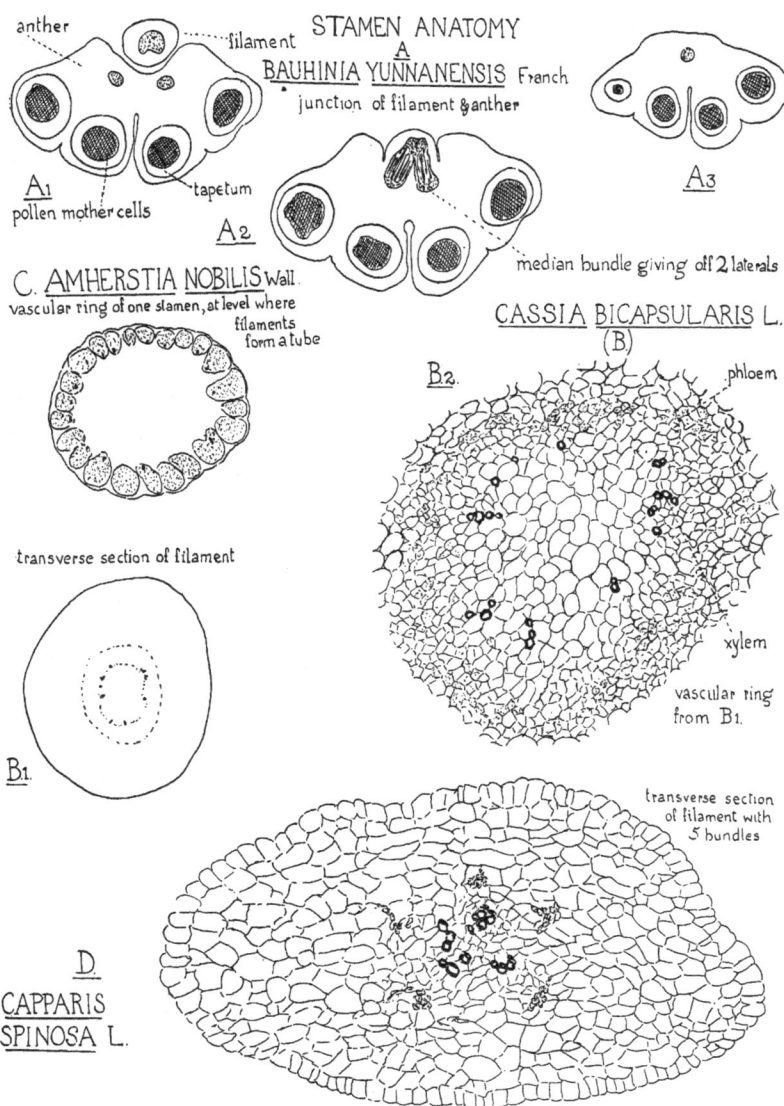

Fig. 5. Stamen anatomy. A 1–A 3, series from below upwards (× 47);
B 1, C (× 47); D (herbarium material) and B 2 (× 193 *circa*).

the ovules, is given by certain abnormal flowers of the clover, *Trifolium repens* L., in which, in place of the ordinary gynaeceum, transitional examples can be found, ranging from members which are simply and obviously foliar, to phyllomes with incurved ovule-bearing margins, approaching the status of the normal legume.

A useful corollary to Goethe and de Candolle's view, is the suggestion that the closest term of comparison for certain carpels is the *peltate* leaf. This idea was put forward in the nineteenth century by Čelakovský, and has been elaborated more recently by Troll.[1]

A feature which sometimes makes the analysis of the gynaeceum less easy, is that certain of its carpels may be imperfectly developed and infertile. In *Triglochin palustre* L., for instance, three of the six carpels are sterile and have no cavity. In gynaecea with fused carpels, an even more extreme stage may be reached, in which the sterile carpels are scarcely recognisable, though it can be demonstrated that they play some small part in the formation of the ovary wall. Vegetative phyllomes, when reduced to the last extreme, may be non-vascular, and the same is true of certain carpels.[2]

Despite the undoubted general success of Goethe and de Candolle's theory of the gynaeceum, one is left with the feeling that there is something unsatisfying in the insistence on a *direct* homology between foliage-leaves and carpels. The Goethean conception is bound up with the idea that the phyllome is an organ *sui generis*. At a later stage[3] we will consider the possibility of replacing this notion by an interpretation of the leaf, which sets it in closer relation to the rest of the plant, and which may enable us to form a picture of the flower including but transcending the Goethe-de Candolle theory.

[1] For Troll's work on the subject, and for references to Čelakovský, see Troll, W. (1932) and (1934*b*); for an anatomical comparison of carpel and foliage-leaf, with special reference to Troll's theory, see Sprotte, K. (1940). Troll's general theory of the gynaeceum will be found in a series of papers in *Planta* (1928*b*, 1931–3); (1935, etc.), vol. I, pp. 14–15; for a criticism of Troll's view, see Arber, A. (1942).

[2] For an important account of pseudo-monomery, with full references to the literature, see Eckardt, T. (1937).

[3] See pp. 70 *et seq.*

CHAPTER V

THE CONCEPT OF THE ORGANISATION TYPE[1]

GOETHE's whole conception of plant structure, which we have considered in the preceding chapter, is permeated by the idea of the type. He offered, as an intellectual basis for vegetable morphology, his notion of the prototype plant (*Urpflanze*)— the common idea (*Begriff*) under which all plant forms might be brought together. Though the type concept is implicit in his essay of 1790, the emphasis there is on the appendages, rather than on the plant as a whole, and he did not explicitly refer to the *Urpflanze* until 1817. He says, in his botanical autobiography, that this concept hovered before him in sensible form, though he recognised that it was actually supersensible;[2] elsewhere he calls it, "eine symbolische Pflanze".[3] If we translate *Urpflanze* as 'primitive plant', or 'primaeval plant', we are reading into it an evolutionary meaning which would have been foreign to Goethe's mind. To him the *Urpflanze* was a *concept*, from which the concepts of existing plant forms could be derived mentally; it carried no phylogenetic implications, and did not to him suggest any notion of an ancestral stock.

Similar ideas to those of Goethe were reached independently by his contemporary, Joseph-François Corréa da Serra (1750–1823), a botanist whose name seems to have passed into undeserved oblivion. Corréa da Serra was Portuguese by birth. He received a good education, and in youth he studied in Italy and sojourned for a long time in Rome. Being recalled to Portugal, he was instrumental in founding the Academy of Sciences in Lisbon. In 1786 he was so unfortunate as to be denounced by the Inquisition, but he found an asylum in France. Later he was able to return to his own country, but other troubles

---

[1] By permission of the Editor of *Biological Reviews*, parts of Arber, A. (1937a) have been incorporated in this chapter and elsewhere.

[2] Troll, W. (1926), p. 202.

[3] Troll, W. (1926) [Goethe, J. W. von, *Glückliches Ereignis*, p. 267].

followed, and he was obliged to seek refuge in England, where he was received cordially by Sir Joseph Banks and the Royal Society. It may be recalled that it was he who brought Robert Brown to Banks's notice, and initiated the friendship and collaboration with Banks which became Brown's mainstay. Towards the end of Corréa's life, he travelled and lectured in America. His experience of men and countries was thus wide and varied. D'Almeida,[1] who is the authority for the facts of his life, points out that Corréa da Serra's existence was 'trop orageuse' to allow of any large-scale work; it may be, indeed, that his stormy career was responsible for the apparent 'paresse insouciante', of which de Candolle accuses him.[2] The little that he published was, however, of an original character, and disclosed luminous ideas. Though his theoretical views were set forth merely by the way, in the course of small taxonomic and descriptive papers, they were of the most far-seeing kind, and expressed with a vivid literary sense, which makes one realise how it came about that his conversational *mots* delighted Paris; one which has survived is his description of the bizarre vegetation of New Holland as "Flore au bal masqué". He was just the man to throw out hints and suggestions, which more pedestrian minds could develop in detail. He affected the stream of morphological thought mainly through the influence which he exerted on A. P. de Candolle. The principal idea which de Candolle took from him was that of the plan underlying each group—a plan "suivi avec ténacité, mais varié avec richesse",[3] and enforced with "stubborn versatility".[4] This idea is echoed in de Candolle's *Théorie Élémentaire de la Botanique*, which is not a mere text-book, but an attempt to expound in their generality the logical principles which should serve as the basis for the study of organised beings. In this work de Candolle reached the point of centring his morphological interpretations in the concept of 'types primitifs et réguliers', of which he regarded all existing irregular forms as modifications.[5] At a later stage he elaborated this idea, stating that "chaque famille de plantes, comme chaque classe de cristaux, peut être réprésentée par un état régulier, tantôt visible par les yeux,

---

[1] Almeida, F. d' (1824); see also Candolle, A. P. de (1862), pp. 162–4.
[2] Candolle, A. P. de (1862), p. 414.   [3] Corréa da Serra, J. F. (1806), p. 62.
[4] Corréa da Serra, J. F. (1796), p. 494.   [5] Candolle, A. P. de (1813), p. 144.

tantôt concevable par l'intelligence; c'est ce que j'appelle son *type*".[1] Though neither Goethe, Corréa da Serra, nor de Candolle appear to have realised the fact, the roots of this conception can be traced back as far as Aristotle, who points out that, within any given group of animals, "the parts are identical save only for a difference in the way of excess or defect".[2] Aristotle's expression—"excess or defect"—implies deviations from an intellectually conceived type form, which is the norm or standard. His hint, pregnant as it was, remained undeveloped by later thinkers, and it is to the independent advocacy of Goethe and de Candolle that the type concept owed the part which it played in the morphology of the nineteenth century, and its still more important role in the modern revival of this branch of botany at the hands of Wilhelm Troll and his pupils.

Reviewing the type theory as expounded by Goethe, we see that he treated various issues together, which, for lucidity, it is well to keep apart. Within the plant itself, he postulated a single type for the different lateral appendages of the stem (e.g. foliage-leaves and flower parts) and other types of higher orders for different kinds of shoot (e.g. leafy shoots and flowers; or expanded inflorescences and capitula). Turning to the plant as a whole, he hinted at the concept of a type for each family of plants—an idea which de Candolle developed in full, on the background of his wide taxonomic knowledge. Finally Goethe postulated an archetype for all flowering plants—the *Urpflanze*.

When we attempt to consider the archetypal plant, we are at once confronted with the main difficulty in the employment of the type concept, namely, that the mind has an almost irresistible propensity to transfer it from the plane of abstractions, where it belongs, to that of sensuous thinking—the plane of the visible and tangible. Goethe definitely regarded his archetypal plant as a supersensible conception, but he perhaps hardly realised how easily one might slip into the error of thinking about it pictorially, while believing oneself to be approaching it abstractly. This difficulty came to the fore in a certain historic botanical discussion which he held with Schiller.[3] Goethe tells us that he demon-

---

[1] Candolle, A. P. de (1827), vol. II, p. 241.
[2] Thompson, D'Arcy W. (1910), vol. IV [*Hist. animal.* I. 1. 486a. [Oxf. trans.)].
[3] Troll, W. (1926) [Goethe, J. W. von, *Glückliches Ereignis*, p. 267].

strated his theory of the metamorphosis of plants to Schiller in vivid fashion, and that, with several characteristic strokes of the pen, he caused the symbolic plant to arise before his eyes; but Schiller shook his head, and said, "This is not an experience (keine Erfahrung); it is an idea." Schiller's implication, that he was confusing abstract thought with sensuous perception, was, perhaps, scarcely fair to Goethe, since he was, in fact, feeling his way, even if half unconsciously, to a mode of contemplative thought in which both these activities should be synthesised and transcended.[1] However, the criticism, even if it did not find its target in Goethe, was richly deserved by some of the lesser men who followed in his tracks. Turpin, for instance, attempted in 1804 an elaborate portrait of the archetypal flowering plant; at that date he had not read Goethe's *Metamorphose*, but, in a book more than thirty years later,[2] he published his picture as an appropriate illustration of Goethe's ideas. It appeared as an engraving, the large size of which made it possible to represent the "Végétal type, idéal, appendiculé", as more than 19 inches high; and this space is needed, for every variety of incompatible detail is crowded into it. The unity of the plant is preserved only by its possession of one main axis. This axis bears a whole series of cotyledons of divers shapes, which are succeeded by a bewildering assortment of leaves of every kind—simple and compound, tendrillar, bulbil-bearing, and rooting—associated with a compendium of axillary branches, including such unusual forms as a fertile cladode. A gradation of members from foliage-leaves to stamens is shown in connexion with a terminal flower. This flower, and another on a lateral branch, obligingly offer examples of different forms of gynaeceum, and also show individual anthers dehiscing by distinct mechanisms. The whole thing is a botanist's nightmare, in which features, which could not possibly coexist, are forced into the crudest juxtaposition.

Long after Turpin constructed his picture, another representation of the *Urpflanze* was published by Schleiden.[3] It has the merit of being simpler than that of Turpin, but it belies

---

[1] See pp. 208–11.
[2] Turpin, P. J. F. (1837), pl. 3.
[3] Schleiden, M. J. (1848), pl. 4, and p. 83.

Goethe's conception of the floral members, and would, one feels, have been almost as distressing to him as Turpin's effort. The absurdity of these particular pictures is so patent that it becomes harmless, but the same inclination to give the archetype visible and tangible expression, took, at a later date, a subtler and hence more insidious form.

In the period that opened with the publication of *The Origin of Species*, the scientific world became convinced, both that evolution had taken place, and also that the natural selection of chance variations provided a master key to the understanding of the process.[1] Up to that time plant forms had been considered worthy of study in and for themselves, and where relations between these forms were recognised, this relation was treated as logical rather than temporal. In the Darwinian reorientation of biology, however, the attention of most botanists was diverted from pure morphology to the use of form data in support of speculations about evolution. This was particularly so where flowering plants were concerned, since the most direct kind of evidence, that of the geological record, was rarely available. To evolutionary schemes, the type concept fell an immediate victim.[2] The Darwinian school seized upon Goethe's archetypal flowering plant, and the notion, common to him and de Candolle, of a minor archetype for each family; detached these ideas from their context in the world of thought; set them up in the world of experience; and assumed their actual historic existence. Goethe's conception of the *Urpflanze*, which in his mind had a timeless quality, was thus transferred to some specific period of the past, as the Ancestral Plant; and it was imaged as something which would have been visible and tangible, if mankind had been there to see and handle. In the intellectual atmosphere of the later nineteenth century, this forcing of Goethe's ideas into an evolutionary frame, seemed a perfectly natural proceeding. To many workers at that time, the diversion of biology into historical channels was a welcome relief, since it transformed theoretical botany into something material,

---

[1] See Robson, G. C. and Richards, O. W. (1936) for a critical, but cautious and sympathetic, revaluation of the selection hypothesis.

[2] According to Meyer, A. (1934), pp. 128–30, in zoology, the transformation of typology into phylogenetics can be witnessed in C. Gegenbaur's work; cf. the first (1859, pre-Darwinian) edition of his *Grundzüge der Vergleichenden Anatomie*, with the second (1870) edition.

amenable to picture-thinking, and not demanding difficult mental activity of a metaphysical kind. Thus, by a feat of legerdemain, which seems to have passed almost unnoticed, the Ancestral Plant was substituted for the Archetypal Plant, and those characters which had, with reason, been attributed to the mental conception of the archetype, were, without further justification, assumed to have been proven for an actual, historically existent ancestor. In justice it should, however, be recalled that T. H. Huxley[1]—deeply imbued though he was with Darwinian ideas—had the candour to admit that the existence of morphological relations between species was not actually incompatible with the doctrine of Special Creation, as expounded in Genesis. Such recognition of the independence in thought of morphology and phylogeny—of the logical and the genealogical, to use the terms of a recent French writer[2]—was, however, rare. The abstract nature of the archetypal plant was glossed over by most biologists, and, if a series could be constructed in the mind, leading from one kind of plant to another, it was considered legitimate to endow the intermediate mental constructs with existence in time and space. It was assumed that one organism was the *descendant* of another, when it was proven, or even merely surmised, that it was *subsequent* to that other—the same fallacy that, in argument, identifies *propter hoc* with *post hoc*. This confusion lent facility to the tracing of phylogenies, a pursuit which, for a long time, fascinated and obsessed morphological thought. The manufacture of these pedigrees was greatly simplified by the assumption that the progress of evolution on all the main lines was necessarily from he simple to the complex, so that the series studied were regarded, without any doubt, as having an irreversible direction in this sense; indeed, in the animal world, evolution towards man as a climax was assumed as too self-evident to need any kind of proof. It is curious that this assumption should have been made so lightly, for it appears that, to Plato, it was the opposite hypothesis that seemed the obvious one. According to the myth of creation in the *Timaeus*, birds, land animals, fishes, shell-fish, and so on, do not represent progressive series proceeding upwards to the human type, but are, on the contrary, degradations

---

[1] Huxley, T. H. (1888), p. 112.
[2] Hocquette, M. (1946), p. 121.

from that type.[1] The land animals, for instance, are imaged as having come from men, who, having no use for philosophy, had their forelimbs and heads drawn down to earth.[2] This shows that the idea of a progression in living things from relative simplicity to relative complexity has not always been regarded as an inescapable principle; but it seemed so to nineteenth century biologists, and it ratified their faith in the compilation of schemes of botanical ancestry. No doubt as to the factual existence of these ancestries was allowed to intrude, and the only serious differences of opinion arose out of the question as to whether ontogeny, or teratology, or comparison of mature forms, or other branches of study, provided the most reliable clues to the unravelling of pedigrees. In other words, the existence of a phylogenetic tree was regarded as 'given'; and it only remained to discuss what particular evidence best revealed its ramifications. Gradually, however, the facile Darwinian view—so easy to understand, and therefore so fatally easy to accept—lost its hold, and morphologists began to question, not merely what evidence might be used, but even whether the problem of phylogenetics itself was a genuine or an imaginary one. It was clear that structural series could be made out among living things, but were these series temporal or merely logical? As far as flowering plants were concerned, phylogenetic schemes had been based on the assumption that there was an actual historic 'tree', starting from a single original stock, and that there was no reason why the course of the branches up to the ultimate twigs should not eventually be exposed to view; but the possibility had at last to be faced that no such tree might ever have existed, and that morphological series might be merely mental constructs with no validity in time. This full scepticism was reached slowly and with much hesitation. The first step was the postulation of two or several original stocks, instead of the single one, but gradually polyphylesis[3] on a much more extensive scale suggested itself, until, first a

[1] It should be noted that the word *archetype* (ἀρχέτυπος) is not used by Plato, but can be traced back to Philo Judaeus (*fl. ante* A.D. 40); see Wolfson, H. A. (1947), vol. I, p. 238.

[2] Cornford, F. M. (1937), p. 358.

[3] For early references to polyphylesis see Čelakovský, L. J. (1878), footnote, p. 154. See also Huxley, T. H. (1888), pp. 122–3. For an account of H. Przibram's zoological work, in which the idea of polyphylesis reaches its extremest expression, see Calman, W. T. (1930), pp. 89–90.

V. THE CONCEPT OF THE ORGANISATION TYPE

bush,[1] and at last a sheaf, rather than a tree, came to seem an appropriate metaphor. In any case, the tree analogy cannot be pushed very far, for phylogenetic trees have the peculiarity of being rootless.[2] Finally, as taxonomic and morphological work disclosed vista after vista of complexity, these simple images of tree and sheaf were replaced by the picture of a tangled reticulum in more than two dimensions. This was not altogether a new idea; it may be regarded as an outcome of an opinion expressed in 1810 by Robert Brown, who wrote of Nature herself connecting groups of organisms net-wise rather than chain-wise.[3] This view of the matter has no obvious relation to the orthodox theory of descent with modification, and its complexity converts into a groping pursuit of will-o'-the-wisp, the attempt to deduce the characters of synthetic ancestral stocks. The reticulum picture thus involves a loss of faith in phylogenetics, but this loss has been a loss of shackles, and has given freedom for a revival of comparative morphology, studied for its own sake, and not in subservience to any evolutionary theory. This change of outlook has increased rather than diminished the interest of typology—indeed the modern German school of morphology is disposed to take "Back to Goethe" as its motto. The change has, however, demanded a reassessment of our scheme of thought; we have to decide what exact significance in this reorientation is to be retained by the 'type' concept.[4]

In this connexion it is helpful to consider the views of certain pre-Darwinian writers. In 1768, long before a belief in evolution had become part of the biologist's creed, Robinet[5] published a book in which he advocated the idea that all beings—passing upwards from the stone, through plant, insect, reptile, and quadruped, to the most excellent form of Being, which is 'la forme humaine'—are 'conçus et formés' after a 'model' or 'original

[1] The 'bush' analogy was suggested by Sachs long ago (cf. Pringsheim, E. G. (1932), p. 146); it has been emphasised recently by other writers, e.g. Vialleton, L. (1930), p. 373.

[2] Hayata, B. (1931), p. 332.

[3] Brown, R. (1810), vol. I. Praemonenda: "ipsa natura enim, corpora organica reticulatìm potiùs quam catinatìm connectens".

[4] The relation of evolutionary and typological morphology is discussed in Thomas, H. Hamshaw (1947) from the standpoint of a phylogeneticist.

[5] Robinet, J. B. R. (1768a or 1768b); for a study of Robinet's idea of the type, see Lovejoy, A. O. (1936), especially pp. 278–80.

example', which he calls the 'prototype'. He was thus, in a certain sense, foreshadowing the type concept of later writers. He illustrates his views by pointing out that a cave dwelling, a savage's hut, a shepherd's cot, an ordinary house, and a palace, may all be considered as graduated variations upon one architectural plan. None of the humbler forms claims to be an Escurial or a Louvre, but they all hark back to the same 'dessein primitif' as the most magnificent palace, since every kind of dwelling is a product of one and the same basic idea, developed to a greater or lesser extent. The intensity of Robinet's desire to find transitional connecting links of the Chain of Being, led him into some extravagant beliefs; he refers, for instance, in all seriousness to a merman who was fished out of the sea on the Suffolk coast in A.D. 1187. This unfortunate credulity coexisted in his mind, however, with a capacity for real insight. He used the word 'évolution' in describing such relations as those between the different kinds of human dwelling, but this term did not mean to him an *historic* movement, progressing through time, such as the phylogeneticist visualises. He has, on the contrary, the merit of considering form *in itself*—not as subordinate to time, but *sub specie aeternitatis*. It is to this standpoint that morphology has to readjust itself, if the study of form is to reveal its full significance; there must be a return to considering the type as a purely abstract intellectual concept, bearing the same relation to the individuals making up the group as the concept 'man' (using the word as a 'universal' in the philosopher's sense) bears to the aggregate of individual men. The types or 'primal patterns', as Owen[1] called them, thus have much in common with certain aspects of the Platonic 'forms'. In the comparative study of morphology, the idea of the type offers a fixed centre from which to appraise the structural variations which occur within every group. In 1840, long after Robinet's time, but before the tide of belief in evolution had set in, Whewell had expressed this thought in a peculiarly luminous way. He had the insight to realise that a natural class of objects "is determined, not by a boundary line without, but by a central point within; not by what it strictly excludes, but by what it eminently includes; by an example, not by a precept; in short, instead of Definition, we have a *Type* for our director".[2]

[1] Owen, R. (1894), vol. I, pp. 387–8   [2] Whewell, W. (1840), vol. I, p. 476.

In this passage, Whewell "builded better than he knew", for his further discussion shows that he did not, in fact, rise to an abstract conception of the type itself, but that he identified the type of a genus, for instance, with the species actually existing which showed the most typical characters. Despite the *aperçu* just quoted, he had not become fully aware that the type concept is essentially mental—an intellectual instrument wherewith the mind brings order into the variegated manifold of phenomena.

When we turn from these broader questions to the individual plant and its components, we must consider what, in the eyes of the modern botanist, should be the status of Goethe's type appendage, which he called the *leaf* (*Blatt*), but for which the generalised term, *phyllome*, is preferable. Like the archetypal plant, the type phyllome is a valid concept so long as it is kept consistently on the abstract plane, but this has not always been done. Goethe's disciples have been too apt to speak as though the sepal and the carpel, for instance, owed their derivation to the foliage-leaf: even a botanist so sagacious as Robert Brown, wrote in 1822 of "the Leaf, from whose modifications all the parts of the flower seem to be formed",[1] while, in the twentieth century, Troll slips into the misconception of calling these parts, "Umbildungsformen von Laubblättern".[2] Goethe himself, however, did not fall into this error, nor did the most discriminating of his followers. Asa Gray, for instance, made it clear that it was a mistake to suppose that petals, stamens, and carpels had existed previously in the state of foliage,[3] and he held that the expression, 'metamorphosis', ought to be used in a purely figurative sense.[4] Goethe, indeed, always insisted on relating the phenomena of metamorphosis to one another *in both directions*. He held that it was as legitimate to call a foliage-leaf an expanded sepal, as to call a sepal a contracted foliage-leaf; this reversibility obviously precludes the idea of historic derivation.[5] The metamorphosis theory, as Goethe himself understood it, thus means that the generally recognised relationship between the different appendicular members arises out of the fact that they are all manifestations of one type-phyllome, non-historic in character. This idea has

---

[1] Brown, R. (1822), footnote, p. 211.
[2] Troll, W. (1935, etc.), vol. I, p. 36.   [3] Gray, A. (1858), p. 231.
[4] Gray, A. (1887), p. 169.   [5] Cf. p. 159.

a number of obvious advantages as a working hypothesis; but the question remains whether it is to be received as an ultimate dictum, or whether, in present-day thought, some further and more satisfying generalisation can be developed out of it.[1] We cannot come to any critical conclusion on this point if we concentrate attention exclusively on the phyllome itself; the problem is a wide one, and in the next chapter we will treat it on a less restricted background.

[1] Cf. pp. 85, 86, 159–161.

THE PARTIAL-SHOOT THEORY
OF THE LEAF

U P to this point in our discussion, we have adopted, as a provisional scheme, the view, implicit in most of Goethe's work, that the aerial part of the plant consists of a stem bearing phyllomes of different kinds, which are comparable with one another, but are not of the same category as the stem. This distinctness is emphasised in the name 'appendicular organs', proposed for them by Turpin in 1820.[2] The notion that leaf and stem are ultimate and discrete units of the plant body, is indeed of great antiquity; it was fostered, no doubt, by the observation of autumnal leaf-fall, which was taken, not unnaturally, to indicate an essential discontinuity between the leaf and the axis which bore it. De Candolle regarded root, stem, and leaf, as separate units,[3] and the same view of the plant body has continued to the present day. Troll, for instance, holds that the leaf is a *Grundorgan*, an ultimate 'given' element in the construction of the plant body, and that the root and the stem are equally basic.[4] From the earliest times, however, another belief has existed, side by side with this rather rigid analysis—the belief that the stem-and-leaf complex, or shoot, has a unified existence of its own. The possibility of grafting buds on to an alien stock was one of the points that led botanists long ago to the realisation of the *individuality* of the bud, and hence of the shoot into which it developed. The way in which each branch shoot echoes the characters of the parent shoot is, indeed, too obvious to need labouring. One instance of their parallelism may, however, be recalled—the presence of prophylls, followed by bud-scales, at the bases of many lateral shoots; an arrangement resembling that of the

[1] By permission of the Editor of *Biological Reviews*, certain parts of Arber, A. (1941 a) have been incorporated in this and later chapters.
[2] Turpin, P. J. F. (1820), p. 29.
[3] Candolle, A. P. de (1827), vol. I, pp. 139–40.
[4] Troll, W. (1935 etc.), vol. I, pp. 176 and 961.

cotyledons, followed by simple 'juvenile' leaves, at the base of the primary plumular shoot.[1] Though the treatment of the leaf as an entity *sui generis* is—especially from the descriptive and taxonomic standpoints—too useful a method to be lightly discarded, the idea of the shoot as the unit is—for the fundamental interpretation of the plant body—far more fruitful.[2] Sachs realised this when he emphasised that "*The morphological conceptions of Stem and Leaf are correlative*; one cannot be conceived without the other ...the expressions Stem and Leaf denote only certain relationships of the parts of a whole—the Shoot".[3]

Botanists have analysed the plant body according to a variety of schemes, and absolute 'rightness' or 'wrongness' cannot be claimed as the prerogative of any one of them. Each way of contemplating the plant—provided that it is a genuine and logical attempt to interpret well-attested facts—may have something of its own to offer, and what we need is a synthetic standpoint, combining the advantages of methods of analysis, which are usually treated as antagonistic. As Swift wrote in 1727, bending a classical saying to his own purpose: "Disputing...should be always so managed that the only end of it is Peace"[4]—an aphorism which finds its expression in philosophic terms in the Hegelian principle of the identity of opposites.[5] The conception of the shoot as the ultimate unit, and the opposed conception of the leaf and stem as each being ultimate units, may be taken as an example of an antagonism which seems at first sight irreconcilable, but which reveals itself in a different aspect, when we look at it more closely. The shoot hypothesis undoubtedly fits better with the broad facts, but it disregards the usual sharpness of the distinction between leaf and stem, and their idiosyncratic character. These features, on the other hand, receive full recognition when seen from the standpoint of stem and leaf as fundamental categories, while here it is the nature of the shoot system as a whole

[1] Cf. Troll, W. (1935 etc.), vol. I, p. 447.

[2] For a more detailed consideration of the units making up the aerial parts of the plant body, see Arber, A. (1930).

[3] "Die morphologischen Begriffe Stamm und Blatt sind correlative Begriffe; eines ohne das andere ist nicht denkbar." Sachs, J. von (1870), p. 134 (this passage does not seem to occur in the first edition, 1868). For the translation cited, see Sachs, J. von (1875), p. 136.

[4] Swift, J. (1727); see *Thoughts on Various Subjects*, vol. II, p. 342.

[5] Cf. Stace, W. T. (1924), p. 94, *et passim*.

which is incompletely imaged. We thus see that the conflict between these two views, though it is undeniably a genuine one, arises out of the fact that each, though true, is true only *up to a point*. We should aim at including and transcending both in a synthesis, which, while treating the shoot as a primary unit, will yet have absorbed into itself such truth as is to be found in the concept of the antithesis of stem and leaf.

Those who regard the leaf as a basic entity, consider every kind of appendicular organ as a modification of the type phyllome. Beyond this, their theory does not, however, allow them to go; for the phyllome, being on this view a fundamental element, cannot be analysed further, or derived from anything else.

Those, on the other hand, to whom the shoot is the unit, are free to ask, as Adrian Spieghel[1] did, three hundred years ago, "But what is the Leaf?" At that date he could only conclude, "Long and oft have I sought an answer to this question, without finding anything with which I could satisfy you and myself." We have now to enquire whether the development of botanical thought since Spieghel's day has given us any clue towards an answer to the problem which he had himself no adequate means of solving.

Nehemiah Grew, writing more than sixty years later than Spieghel, got so far as to recognise the substantial identity of the leaf and the branch which bears it. He says "the Skin of the Leaf is only the amplification of that of a *Branch*....The Fibres or Nerves dispersed through the Leaf, are only the Ramifications of the *Branch's* Wood, or *Lignous Body*. The *Parenchyma* of the Leaf...is nothing else but the continuations of the *Cortical Body*".[2] This conclusion, based on the most primitive micro-technique, has been confirmed by modern workers; it has been shown, for instance, with the aid of twentieth-century methods, that the leaf of the apple (*Malus communis* DC.) strikingly resembles a branch system, both in structure and ontogeny.[3] Other indications of the fact that Grew equated leaf and branch are that he calls the midrib the 'middle *Stem*' of the leaf;[4] and that he names

---

[1] Spieghel, A. (1606), p. 7. "Sed quid...Folium? Hoc diù multùmq. quaesivi, nec quidquam inueni, quo mihi vobisq; satisfaciam."

[2] Grew, N. (1672), pp. 109, 110.   [3] Blaser, H. W. & Einset, J. (1948).

[4] Grew, N. (1682), p. 175.

the ultimate segments of the extremely shoot-like leaf of fennel, "a company of Ramulets".[1] Jung, again, pointed out that the elaborate leaves of parsley (*'Apium hortense'* = *Carum Petroselinum* Benth. et Hook.), columbine (*Aquilegia*), and paeony (*Paeonia*), simulate branches;[2] this recalls the description by Albertus Magnus of the rachis, in certain compound-pinnate leaves, as *ramus folialis* or *virgula parvula*.[3]

At a much later date, Goethe, in his study of the relations of the appendages,[4] as a rule treats the leaf as being an organ *sui generis*, and this view alone seems to have become associated with his name, and to have been adopted by his disciples. Certain incidental references show, however, that he occasionally had a fleeting glimpse of the leaf from a less confined standpoint, and that he then shared the vision of those earlier writers whom we have just cited. Speaking of leaves formed of many leaflets, he says that "they prefigure complete small branches",[5] and he also describes compound leaves as "in reality branches, the buds of which cannot develop, since the common stalk is too frail".[6] Moreover, in 1790—the year in which Goethe's *Metamorphose* was published—the philosopher, Immanuel Kant, made the cognate suggestion that "each *branch* or *leaf* of a tree may be regarded as merely engrafted or inoculated upon it, and therefore as a tree with an existence of its own, simply attached to another from which it nourishes itself".[7] He thus implies that leaf and lateral shoot are *parallel members*, each possessing an individuality distinct from that of the parent shoot, and that they are each, in a certain sense, equivalent to the plant as a whole. This conception opens the way to a deeper understanding of plant construction; on the other hand, Goethe's prevailing view that the leaf is a 'given' entity, which defies analysis, puts an effective stop to further discussion.

---

[1] Grew, N. (1672), p. 114.   [2] Jung, J. (1747), cap. iii, 27, 29, pp. 9, 10.
[3] Sprague, T. A. (1933*b*), p. 433.
[4] Goethe, J. W. von (1790), and later notes.
[5] Goethe, J. W. von (1790), §20.
[6] Goethe, J. W. von (1887 etc.), Abt. ii, Bd. 13, Par. 130, p. 125.
[7] "Daher kann man auch an demselben Baume jeden Zweig oder Blatt als bloss auf diesem gefropft oder oculirt, mithin als einen für sich selbst bestehenden Baum, der sich nur an einen andern anhängt und parasitisch nährt, ansehen." Kant, I. (1790), *Kritik der teleologischen Urteilskraft*, quoted from Kant, I. (1902–38), vol. v, 1908, pp. 371–2; see also Meredith, J. C. (1928), p. 19, for another translation.

Kant's remarks seem to have passed unnoticed, though from time to time, during the nineteenth century, views, not unlike his, found expression among botanists. In 1861, for instance, Schultz-Schultzenstein pointed out that to adopt the leaf as an ultimate unit was unsound, since the nature of the leaf was itself a question—"das grösste Räthsel der Botanik"—for which an answer ought to be sought.[1] One of the first advances towards a solution came from Dresser, who, in 1859, threw out the suggestion that the leaf was a modified branch.[2] Nearly a decade later, Casimir de Candolle, A. P. de Candolle's grandson, tackled the problem with scientific thoroughness,[3] and concluded that the ordinary bifacial leaf is a branch, altered by atrophy of the 'cône terminal', both at its apex ('son sommet') and also on the adaxial face, while, in peltate and cylindrical leaves, this atrophy affects the apex alone. De Candolle's idea may be re-expressed by saying that *the leaf is a partial-shoot, arising laterally from a parent whole-shoot.*

This illuminating theory of the leaf, as well as the earlier hints which led up to it, was suggested by observations upon flowering plants, but it is found to fit into the framework of certain interpretations of the plant body in general, which have arisen out of the study of vascular cryptogams, recent and fossil. Students who have concerned themselves primarily with these groups, have not been tempted—like the botanists who deal with angiosperms, and thus have deciduous trees constantly under their eyes—to treat stem and leaf as discrete and comparable units. Before the end of the nineteenth century, a palaeobotanist, Potonié, came to the conclusion that in the ferns there is no absolute heterogeneity between stem and leaf.[4] Tansley, also, based his study of the filicinean vascular system upon the view that "the fern-leaf is in phylogenetic origin a branch or rather a branch-system...and not an appendicular organ differing *ab initio* from the axis on which it is borne".[5] This view has been accepted and developed, especially by those concerned with fossil plants. It has been said, for instance, of *Dichophyllum*, *Psygmophyllum*, and certain

[1] Schultz-Schultzenstein, K. H. von (1861), p. 275.
[2] Dresser, C. (1859), p. 111; see also Schultz-Schultzenstein, K. H. von (1861).
[3] Candolle, A. Casimir P. de (1868).
[4] See, for instance, Potonié, H. (1899), p. 125.
Tansley, A. G. (1908), p. 1.

coenopterid ferns, that "In so far as we know these forms, it is not possible to relegate their branch systems to the classical categories of stem or leaf, and it is probable that the term shoot will be a generally acceptable one."[1] Moreover, as an example from the gymnosperms, it may be recalled that the needles of *Pinus* have been interpreted as 'fragmentary shoots'.[2] The partial-shoot theory of the angiosperm leaf may be regarded as in some sense a corollary of such views as these.

Certain fossil plants of great antiquity, the Rhyniaceae, have provided the starting point for a theory which may be held to include the branch-system conception of the fern leaf, and also to cover a wider field; this is the *telome theory* of Zimmermann.[3] It has been claimed that the Rhyniaceae, which did not become fully known to botanists until the present century, reveal many of the characters of the primeval ancestral stock from which the great divisions of the vegetable kingdom may have been derived.[4] The evidence for this view does not lie within our province, but we must touch upon the telome theory, which has been its outcome, since the immense range of plant forms, to which its originator, Zimmermann, relates it, includes the angiosperms. According to this writer, the primary units, of which the plant body is constructed, are one-bundled ultimate branches of the shoot, which he calls telomes. Telomes may be sterile (phylloid) or fertile (sporangial). He thinks that the sterile and fertile telomes are homologous, and that whether they become vegetative or reproductive depends upon conditions. The shoot axis of the higher plants he regards as developed from the basal parts of telome clusters; leaves are formed either from a single sterile telome, or from a complex of such structures. We need not follow the elaboration of Zimmermann's theory, which he works out meticulously; it is only in its most general aspect that it concerns us here. An essential element in the theory is that leaves are to be conceived as primarily radial in structure, and branched in all directions, and that flattening and dorsiventrality are secondary developments. If these views were expressed in morphological

---

[1] Andrews, H. N. (1941); cf. pp. 377–8.
[2] Dufrenoy, J. (1918), p. 439.
[3] Zimmermann, W. (1930); see, for instance, pp. 65 *et seq.*
[4] Zimmermann, W. (1930), p. 104.

rather than in phylogenetic terms, they would in no way conflict with the partial-shoot theory of the leaf.

When considering the work of Goethe and de Candolle, we found that the evidence pointed to a parallel interpretation for vegetative and floral members.[1] Without reiterating this evidence, we may then conclude that, if the partial-shoot theory is applicable to the foliage-leaf and its variants, it should also be the key to the understanding of sepals, petals, stamens, and carpels. As soon, indeed, as we attempt to use it, we find that it justifies itself by resolving the antithesis between Goethe's idea of the parts of the flower, and certain more modern views. A recent writer,[2] for instance, impressed by the obvious differences between the stamen and the foliage-leaf, has rejected Goethe's picture, and has put forward an explanation of both stamen and carpel, based on the telome theory;[3] that is to say, he regards both these members as 'reduced branch systems'. However, if the partial-shoot theory of the phyllome be adopted, it annuls the main elements of the antagonism between such views[4] and the Goethe-de Candolle interpretation, since, in the flowering plants, not only all kinds of foliage-leaves, bracts and sepals, but also petals, stamens, and carpels, are regarded as partial-shoots, and thus may all be called branch systems, though of an incomplete kind.

As we have seen, the partial-shoot theory supplies a key to most of the main characteristics of phyllomes, but it is not, as Casimir de Candolle left it, wholly satisfying, and it is considerations of a broad and general sort which may help to complete it. Returning to the Greeks, we may recall that the Stoics[5] recognised *self-maintenance* as a fundamental character of living things, and this idea became part of the Aristotelian tradition; for instance, in the pseudo-Aristotelian *De plantis* of Nicolaus Damascenus, who flourished at about the time of the birth of Christ, we read that, "when some form of plant is created, it persists in its own constitution".[6] Boethius (c. A.D. 480–524), who may be regarded as the last of the classical or the first of the medieval

---

[1] See pp. 47–58.   [2] Wilson, C. L. (1942).   [3] See p. 75.
[4] For other non-Goethean theories relating to the flower, see Thomas, H. Hamshaw (1932), and earlier and later papers by this writer.
[5] Pollock, F. (1899), footnote, pp. 109–10.
[6] Cf. Hett, W. S. (1936), p. 163; and Forster, E. S. (1913), 818b.

philosophers, emphasised this principle, and went so far as to apply it specifically to plants. His dialogue with Philosophy contains the passage: "If (quoth I) I consider living creatures, which have any nature to will and to nill, I find nothing that without externe compulsion, forsake the intention to remain, and of their owne accord hasten to distruction....But what I should think of hearbs, and trees,...I am altogether doubtfull." In her reply Philosophy indicates that he ought in no way to doubt that the principle of self-continuance is implicit also in the vegetable world.[1]

In the seventeenth century, Baruch de Spinoza, in the *Short Treatise*,[2] which may be his earliest work, adopted, perhaps through Boethius,[3] the idea of self-maintenance. In the *Ethic*, which represents the full maturity of his thought, he added to the self-maintenance thesis—which is in itself almost a truism— a final touch which transformed it into something of illuminating significance. He gave it the following formulation: "The effort by which each thing endeavours to persevere in its own being is nothing but the actual essence of the thing itself";[4] moreover, at an earlier date, he had definitely identified it with *life*.[5] He thus set the principle in the forefront of biology by making the urge towards self-maintenance, not merely a character of living things, but the very gist of life itself. This urge to self-continuance is revealed in different ways in different living creatures; in the flowering plants, with their unlimited power of growth from fresh centres, it is expressed in repetitive branching. The whole plant may be said to consist of a series of shoot generations, together with a series of root generations; every individual lateral branch

---

[1] Boethius, A. M. T. S. (1609), bk. III, xi, pp. 78 *et seq.*; for a revised version of this translation, see Stewart, H. F. and Rand, E. K. (1918), *The Consolation of Philosophy*, pp. 278–80.

[2] Translated in Wolf, A. (1910), see chap. V, p. 47.   [3] Cf. Arber, A. (1943).

[4] "Conatus, quo unaquaeque res in suo esse perseverare conatur, nihil est praeter ipsius rei actualem essentiam." [Spinoza, B. de] "B. D. S." (1677), Ethices, pars III, Prop. vii, p. 102. For translation see White, W. Hale, and Stirling, A. H. (1930). In the earlier *Tractatus Theologico-Politicus* (cap. XVI) he had described it as "the highest law of Nature" (lex summa Naturae), Vloten, J. van and Land, J. P. N. (1882), vol. I, p. 552. On Spinoza's idea of the *conatus* see also Bidney, D. (1940), pp. 86–112.

[5] "Quare nos per *vitam* intelligimus *vim, per quam res in suo esse perseverant*." *Cogitata metaphysica*, pars. II, cap. VI, in Vloten, J. van, and Land, J. P. N. (1883), vol. II, p. 487.

is a repetition, modified in varying degrees, of the original primary shoot, while every lateral root similarly repeats the primary root. The plant in endeavouring "to persevere in its own being", repeats that being time after time, each daughter shoot or root becoming, in its turn, a parent shoot or root. This process grades into actual reproduction, when the lateral shoots take, for instance, such forms as detachable rooting bulbils. Self-maintenance, indeed, is essentially one with reproduction, which, as Aristotle says, is the only way in which the plant can "share in the immortal and divine...; what persists is not the individual itself, but something in its image, not identical as a unit, but identical in form".[1]

When Spinoza spoke of the organism's persistence in its being, he was thinking of the individual members, as well as of the creature as an entirety; concerning man, he speaks of the way in which each separate limb, considered in itself as a whole, and not merely as a part of the whole organism, strives to maintain its own well-being.[2] Moreover, "the endeavour to persevere in its being" is a phrase richer in significance than may be realised at first glance; it implies, not only continuance in time of the being in question, but also its endeavour to bring its *determinata natura* into a condition of completeness and perfection in its own kind, or, in Aristotelian terms, to actualise its potentialities.[3]

These general ideas, when applied to the problem of the leaf, lead us to supplement the hypothesis, that this member may be interpreted as a partial-shoot, by the corollary, that *the partial-shoot has an inherent urge towards the development of whole-shoot characters*. This view has no phylogenetic implications; it does not commit us to any opinion as to the origin of the leaf as a matter of history, but is concerned with what the leaf actually *is*, here and now. There is no need to assume that the leaf passed ancestrally through a complete-shoot phase, and was subsequently reduced to its present incomplete form: it is in its very essence partial, and this partialness expresses itself as dorsiventrality, complicated by an innate impulse towards radiality.

So much for *the nature of the hypothesis* which we are offering; we must now consider *the evidence*, firstly, for regarding the leaf as

[1] Hett, W. S. (1935), pp. 84–7 [*On the Soul*, II. iv. 415 a].
[2] Wolf, A. (1910), chap. v, p. 47.   [3] Cf. Russell, E. S. (1945), p. 191.

a partial-shoot, and, secondly, for the corollary that this partial-shoot has an inherent urge towards whole-shoot characters. It is convenient, where we can, to keep these aspects apart, but often they are too closely involved for separation.

The ontogenetic history of the leaf is consistent with the partial-shoot theory, for leaf growth is initially terminal, like that of the shoot; but the meristematic activity of the leaf-tip soon comes to an end, and the focus of cell division is transferred to the lower regions. Perhaps the precocious maturity, and the short life of the leaf, may also be symptoms of its imperfect character as a partial-shoot.

When we turn from the question of apical growth to that of lamina development, we find that in dicotyledons this is predominantly pleuroplastic, i.e. the blade arises from the lateral margins of the leaf primordium. This was noticed and illustrated long ago by Trécul;[1] for a more modern account, we may recall Avery's description of foliar development in tobacco (*Nicotiana Tabacum* L.).[2] In this plant the leaf is little more than an embryonic midrib until it reaches 0·6 mm., or more, in length. The lamina then makes its first appearance as two ridge-like projections, initiated by the activity of a row of sub-epidermal cells, forming a marginal meristem, which extends along either side of the midrib primordium. This process seems, at first sight, very different from the ontogeny of the shoot, but it has been shown[3] that, in the shoot apex, the superficial layers grow as a lamellar meristem, and this produces folds which form the leaves. It follows that the leaf-lamina arises from the peg-like rudiment of the leaf in much the same way as the leaves themselves grow from the shoot axis, allowing for the difference between a dorsiventral and a radial system. The petiole may be regarded as part of the leaf in which lamina development is inhibited,[4] just as the naked stem is a part of the shoot in which leaf development is in abeyance.

When we scrutinise the lamina more closely, we see that it arises predominantly along the lateral lines of intersection of the upper and lower surfaces of the phyllome. The marginal relation

[1] Trécul, A. (1853); e.g. pl. 21, figs. 46–52.
[2] Avery, C. S. (1933), pp. 568–9.   [3] Schüepp, O. (1938).
[4] Troll, W. (1932), pp. 178–9.

of lamina to rachis is indicated for a species of *Pterocarya* in Fig. 19, A, p. 117, and for *Lapsana communis* L. in Fig. 18, C 1, C 2, p. 115. In *Melianthus major* L. (Fig. 18, D 1 and D 2) the rachis tends towards a unifacial type, the upper surface being reduced to a bare minimum, so that the lines, along which the rachis-wings arise, are almost in contact (cf. also Fig. 19, A 1, A 3, A 4, p. 117). The origin of the lamina at the junction of the leaf surfaces suggests that there is a significant individualisation of the two faces—the lower, or abaxial, which is a direct continuation of the shoot surface, and the upper, or adaxial, which may be visualised as a fold in the shoot surface, and thus as more internalised than the outer surface. The existence of this difference is suggested from the very first by certain cell features. It has been demonstrated, for instance, in the *Phlox*,[1] that, in each foliar primordium, vacuolation of the cells occurs first on the abaxial face of the procambial strand, and later, on the adaxial face.

If we consider the leaf as a shoot 'in little', the elements in its construction which may be compared more particularly with the axial or stem elements of the shoot, are: the median region of the leaf-base; the petiole; the midrib of the simple leaf; the rachis and racheoles of the compound leaf; the non-winged leaf apex of certain leaves, and the rachis-tendril of others. The elements of the leaf, on the other hand, which parallel the foliar elements of the shoot are: the stipules; the lateral wings of the sheathing leaf-base; the lamina of the simple leaf; and the laminae of the pinnae of various orders forming the compound leaf. It is indeed obvious that, in compound leaves, the relation of the leaflets to the petiole or rachis corresponds with the relation of leaves to the stem which bears them. Sometimes the leaflets fall, leaving the petiole still in place, like an axis surviving the loss of its leaves; this is shown for *Rubus laciniatus* Willd. in Fig. 6, C 1, p. 81. It has been noticed in the horse-chestnut (*Aesculus Hippocastanum* L.) that the absciss layer, proper to the petiole foot, is repeated in every detail at the bases of the petiolules.[2] Stipules, since they have much in common with leaflets, may be considered in the same connexion. The relation of the stipules to the leaves of which they are members, may be compared with the relation of cotyledons to the primary shoot,

[1] Miller, H. A. and Wetmore, R. H. (1946), p. 9.   [2] Leavitt, R. G. (1909), p. 31.

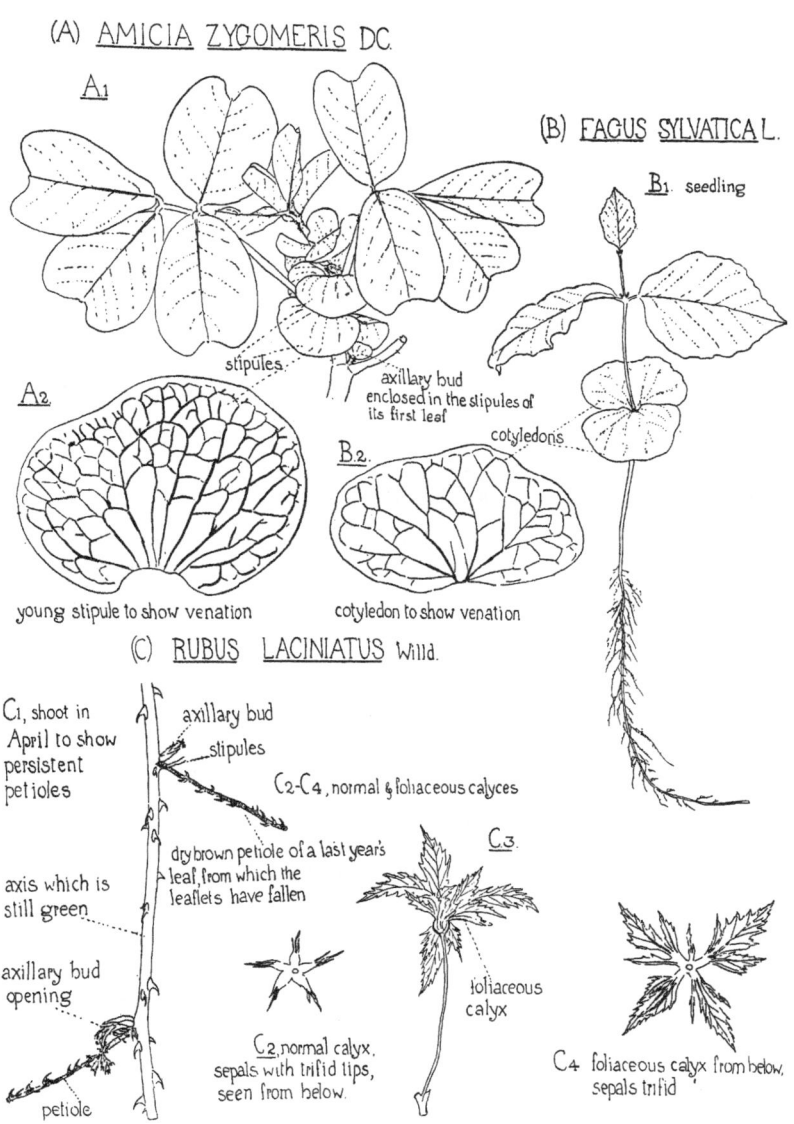

(A) AMICIA ZYGOMERIS DC.

A1

(B) FAGUS SYLVATICA L.

B1. seedling

stipules

axillary bud
enclosed in the stipules of
its first leaf

A2.

cotyledons

B.2.

young stipule to show venation

cotyledon to show venation

(C) RUBUS LACINIATUS Willd.

C1, shoot in
April to show
persistent
petioles

axillary bud

stipules

C2-C4, normal & foliaceous calyces

dry brown petiole of a last year's
leaf, from which the
leaflets have fallen

C3.

axis which is
still green

axillary bud
opening

foliaceous
calyx

C2, normal calyx,
sepals with trifid tips,
seen from below.

C4 foliaceous calyx from below,
sepals trifid

petiole

Fig. 6. A1, B1, C1–C4 ($\times \frac{1}{3}$); A2 ($\times 9$ *circa*); B2 ($\times \frac{5}{6}$ *circa*).

and of prophylls or bracteoles to a lateral shoot. Axillary stipules
—such, for example, as those of *Melianthus major* L. (Fig. 18,
D 1 and D 3, p. 115) and of *Houttuynia cordata* Thunb. (Fig. 27,
B 1 and B 2, p. 146)—may perhaps fall into one category with the
single seed-leaf, prophyll, or bracteole of monocotyledons, while
lateral stipules suggest the same members, but in their paired
dicotyledonous condition. Stipules often differ in form, size,
texture, etc., from the succeeding elements of the leaf (leaflets),
while prophylls and cotyledons are distinguished, correspondingly from the elements of the shoot (foliage-leaves) which
follow them. Individual cotyledons and stipules may, indeed,
bear a decided resemblance; for instance, the broad stipules of
Amicia Zygomeris DC. (Fig. 6, A 1 and A 2, p. 81) suggest in
shape and venation the seed-leaves of *Fagus sylvatica* L., the beech
(Fig. 6, B 1 and B 2). It may be recalled also that, in a certain
detail of minor importance, prophylls and stipules agree—
namely, in their occasional liability to asymmetry. In this respect
the stipules of the hawthorn (Fig. 2, A 3, p. 52) resemble the
prophylls of the potato;[1] these prophylls, which are borne at the
very base of the lateral shoot, might, at first glance, be mistaken
for stipules belonging to the axillant leaf.

Not only can shoot and leaf be considered in serial filiation—
the leaf being regarded as a reduced offspring, which partially
repeats the characters of the shoot—but the leaf, also, may sometimes bear outgrowths which are affiliated to it as is the leaf itself
to the parent shoot. Such outgrowths are not uncommon on
cabbage leaves. The relation between these so-called enations
and the foliage-leaf which produces them; or between the nectary-
scale of *Ranunculus* and the honey-leaf;[2] or between the corona of
Narcissus and the perianth;[3] may be compared with the relation
between a leaf and its parent axis. That there is indeed a basic
similarity between these relations, is confirmed by the anatomy;
we find that leaf-enation, honey-gland, and corona, all have the
xylem-faces of their bundles turned towards the parent member.

It is not only in its external form, and in its relation to members
of the next generation, that the leaf may be compared with the
shoot; the resemblance extends also to the nervation, which

[1] Troll, W. (1935, etc.), pp. 383–4, and fig. 294.
[2] Arber, A. (1936).   [3] Arber, A. (1937*b*).

reveals a definite correspondence with the general shoot-branching of the plant. Uittien,[1] developing a suggestion made by Velenovský with reference to trees,[2] has shown that leaves tend to be pinnately nerved or divided, when the main axis of the shoot in the same species exceeds the lateral axes in length; and palmately nerved or divided, when the lateral axes exceed the main axis. He has worked out such parallelisms between shoots and leaves in detail, and has demonstrated, for instance, that, among families with racemose inflorescences, those in which the flower groups are particularly large and complex, have also, in general, pinnate-compound leaves. Sachs had already realised the parallelism of shoot and leaf, for he not only pointed out that the principles of shoot and leaf branching are the same,[3] but, in the English version of his text-book, the words 'racemose' and 'cymose' are used for types of leaf venation.[4] He had not, however, reached Uittien's generalisation, though Oken, many years before, had given a sign that he had faintly foreshadowed it. Oken wrote: "From the arrangement of the leaf-nerves one can deduce the structure of the whole plant and determine its character."[5] There is reason to believe that Oken's idea may be applicable over a wide field; it might, for instance, be suggested that there is a correlation between the leaf venation of monocotyledons—which shows, as a rule, a less elaborate system of branching than that of dicotyledons—and the relatively unbranched habit, which characterises many monocotyledonous shoots.

The analogy between leaf and shoot has been obscured by the technique of leaf description, which is based on the idea of the leaf as a member with an entire margin, which may be more or less indented or deeply cut, as if a pair of scissors had been employed upon it. De Candolle long ago pointed out that this

---

[1] Uittien, H. (1928 a).

[2] Velenovský, J. (1905–13), vol. II, 1907, p. 626, "der Verzweigungstypus der Krone auch in der Nervatur der Blätter ausgeprägt zu sein pflegt".

[3] Sachs, J. von (1868), p. 158.　　　　[4] Sachs, J. von (1875), p. 162.

[5] "Aus der Anordnung der Blattrippen kann man daher den Bau der ganzen Pflanze erkennen, und ihren Charakter bestimmen.' Oken, L. (1810), vol. II, § 1146, p. 72; for a translation from a later edition see Oken, L. (1847), § 1135, p. 225. The same idea was followed out at a later date by M'Cosh, who compared the 'angle of venation' and the 'angle of ramification' in the same plant; see M'Cosh, J. and Dickie, G. (1856), p. 116.

method of visualising lamina-form is liable to create a wrong impression. He states that "les feuilles ne sont point des surfaces entières qui se découpent; ce sont des portions de limbes qui, en se soudant ou en restant soudées à divers degrés, constituent tantôt des angles saillans ou rentrans, tantôt des surfaces entières."[1] On this view, De Candolle would describe a pinnatifid leaf as showing fusion, for half their length, of the lobes associated with the pinnate lateral veins.[2] He goes so far as to suggest that this theory may possibly be applied to all leaves, even those that are quite simple and undivided; "Mais les feuilles qu'on appelle simples ne seraient-elles autre chose que des feuilles composées à folioles soudées?...Cette théorie est surtout applicable aux feuilles palminerves et peltinerves, qui ne semblent formées que par la soudure de plusieurs folioles palmées ou peltées; mais comme toutes ces folioles ont le limbe penninerve, il en résulte en définitive cette loi remarquable, que toutes les feuilles des dicotylédones pourront un jour être considérées comme les limbes pennés diversement soudés entre eux."[3] De Candolle's view is not a description of the ontogeny, since the mature form is, in fact, the outcome neither of 'fusion' nor of 'cutting'; but, from our present standpoint this is immaterial; we are not concerned with the mechanism of development, but with the intrinsic relation of parts in which it culminates. The point that matters to us is that de Candolle saw, even if dimly, that the leaf is a system comparable with a shoot, but in which the main and lateral veins and their associated leaf surfaces form a united whole, instead of being separable entities, such as the main axes and the lateral branches with their individual leaves, which in the aggregate make up the shoot.

Up to this point we have been concerned primarily with the comparison between the leaf and the shoot. It remains to see how the point of view regarding the nature of the leaf, which we have now reached, affects comparisons, not between leaf and shoot, but—within the leaf itself—between laminae of various kinds.

The prevalent mode of ontogeny among dicotyledonous leaves, is, as we have already noticed, the pleuroplastic, in which a rod-

[1] Candolle, A. P. de (1827), vol. I, p. 299.
[2] Candolle, A. P. de (1827), vol. I, p. 301.
[3] Candolle, A. P. de (1827), vol. I, p. 318.

like rudiment gives rise to the flat lamina by growth from a marginal meristem, thus producing an extended dorsiventral member by lateral winging. In certain dicotyledons, however, the structure diverges widely from this type. In the so-called 'midrib-leaf', or 'rachis-leaf'[1]—for instance, in *Catananche* (Fig. 22, C 1 and C 2, p. 122), *Plantago* (Fig. 22, A), *Scorzonera*, and some species of *Eryngium* (e.g. Fig. 22, B)—the rod-like pre-laminar leaf-rudiment, instead of developing wings, expands laterally, thus separating the veins. There is nothing in such leaves, except perhaps a narrow edging, which is equivalent to the lamina wings of an ordinary pleuroplastic leaf. Among the monocotyledons, again, blades still more unlike that of the pleuroplastic phyllome in origin, are the rule rather than the exception.[2] The leaf of the banana (*Musa sapientum* L.)[3] offers an interesting divergence from the pleuroplastic scheme. In this plant the lamina-halves do not begin to develop until a comparatively late stage, when the marginal tissue of the rudiment has lost its meristematic character, and is thus incapable of serving as a source for wing-development. The meristematic tissue forming the wing has therefore to be supplied by cells farther from the edge; the result is that the lamina-halves developed from this meristem lie, not only within the scarious border, but even within the marginal bundles.

Though the leaf of *Musa* actually diverges so much from the ordinary pleuroplastic leaf, the difference can only be detected on careful inspection, but there are other monocotyledonous genera in which lamina anomalies are much more obvious. In certain Iridoideae, for instance, the leaf is 'foliated'—a vascular wing being associated with each main bundle. *Cypella Herberti* Herb. is a conspicuous example;[4] here the lamina, with its elaborate system of keels, is a very different thing from an ordinary leaf-blade.

These instances show that, in any sound theory of the leaf, the fact must be recognised that the mature lamina may be developed

---

[1] Gaisberg, E. von (1922); Troll, W. (1935 etc.), vol. I, pp. 1630–44.

[2] The writer formerly called the kind of blade met with in the monocotyledons a 'pseudo-lamina' (Arber, A., 1918, 1922 b, etc.); it seems best, however, to drop this term, which is invalidated when the ordinary type of monocotyledonous leaf is, as we shall show on p. 108, no longer to be interpreted as a petiolar phyllode.

[3] Skutch, A. F. (1930).

[4] Arber, A. (1921), fig. 50, N, p. 321; or (1925), fig. xcvi, 50 N, p. 122.

by a variety of methods from a variety of structural elements, and therefore cannot always be tidily homologised from family to family. Now those who adopt a naïve form of the type concept as the basis of their morphology, take it for granted that all leaves are referable, ultimately, to one type; that is to say, they consider them all as amenable to analysis on a single unified scheme, and as thus having a point-for-point correspondence. This opinion follows logically if it is held that the leaf is a prime morphological category. On our present theory, however, this member is no longer regarded as a 'given' and basic entity, so we are relieved from the necessity of forcing all phyllomes, however intrinsically different in their structure, into a framework of homology. As an example we may take a simple, leaf-base-like scale-leaf. In order to make this fit neatly into the type concept of the phyllome, there has been a tendency in the past to homologise it with the leaf-base alone of a fully developed foliage-leaf. Modern ontogenetic work has shown, however, that it cannot be thus equated with one individual component of a completely elaborated foliage-leaf.[1] The real position is thus better expressed by describing scale-leaves and foliage-leaves as parallel members, than by attempting to derive them both from a hypothetical 'type phyllome'. For, according to our view, the need does not arise to visualise a process of leaf evolution, traceable back to a primordial leaf. On the present interpretation, there is no such thing as a primitive phyllome type, from which all the existing forms of foliage-leaves and floral members have been derived, and there is thus no question of equating all these members with such a hypothetical primitive type, and hence with one another. The phyllome can claim no individual lineage *as a phyllome*—from the first it has been neither more nor less than a special phase of the shoot system of the plant to which it belongs, so that its relation to the leaves of other groups is of an indirect kind. On this view phyllomes are not regarded as manifestations of one *type*, but they are treated as *parallel* members, whose shared characteristics depend upon the fact that they are all incomplete shoots.

In considering the leaf as a partial-shoot, there is a certain danger of so far stressing the resemblances between shoot and

---

[1] See the work of A. S. Foster and his school, especially Cross, G. L. (1938), and references there cited.

leaf that full attention may not be given to their differences. One of the most obvious of these differences is that the leaf is dorsiventral, while the shoot is radially symmetrical. Oken was fully alive to the significance of this fact; in his words, "The leaf is a tree of special form, a tree the branches or veins of which all lie in one plane."[1] It is this partial-shoot limitation to two dimensions which conditions the webbing of the branch system forming the leaf, and thus produces a flat assimilating surface. The distinction between the two-dimensional leaf and the three-dimensional shoot is, however, by no means absolute. The tendency to radiality and the tendency to dorsiventrality coexist both in shoot and leaf, though the relative emphasis on these tendencies varies—the shoot being predominantly radial, but with an underlying trend towards dorsiventrality, and the leaf being predominantly dorsiventral, but with an underlying trend towards radiality. Using Aristotelian and Hegelian terms,[2] we might say that in the shoot radiality is actual, or explicit, while dorsiventrality is potential, or implicit; for the leaf, on the other hand, the reverse is true. Both in shoot and leaf, that tendency which is usually latent may become externalised. For instance, in such plants as the grasses, with their two-ranked leaves, the whole vegetative shoot is built on a dorsiventral rather than a radial plan; this dorsiventrality may find further expression in the fertile shoot.[3] In the species of *Paspalum* drawn in Fig. 7, C 1 and C 2, p. 88, for instance, the branches of the inflorescence are flattened and parallel-veined; they look like narrow lanceolate leaves. The spikelets are given off from a thickened median rib on the abaxial face of the branch. In the Leguminous *Acacia alata* R. Br., again, the shoot, with its distichous leaves, of which the bases wing the stem in a vertical plane, has a dorsiventral form, which may be compared with that of certain leaves (Fig. 7, A, p. 88). In such an example as this, the resemblance to a leaf is not so close as to arouse any doubt as to the category to which the organ belongs, but in other cases a shoot may masquerade as a phyllome so effectually that its true nature is scarcely detectable.

[1] "Das Blatt ist ein Baum von besonderer Form, ein Baum dessen Aeste oder Faserbündel alle in einer Ebene liegen." Oken, L. (1810), vol. II, § 1144, p. 72; for a translation from a later edition, see Oken, L. (1847), § 1133, p. 224.

[2] Cf. Stace, W. T. (1924), p. 23.

[3] Arber, A. (1934), pp. 110, 134, 136, 141, 315, 321.

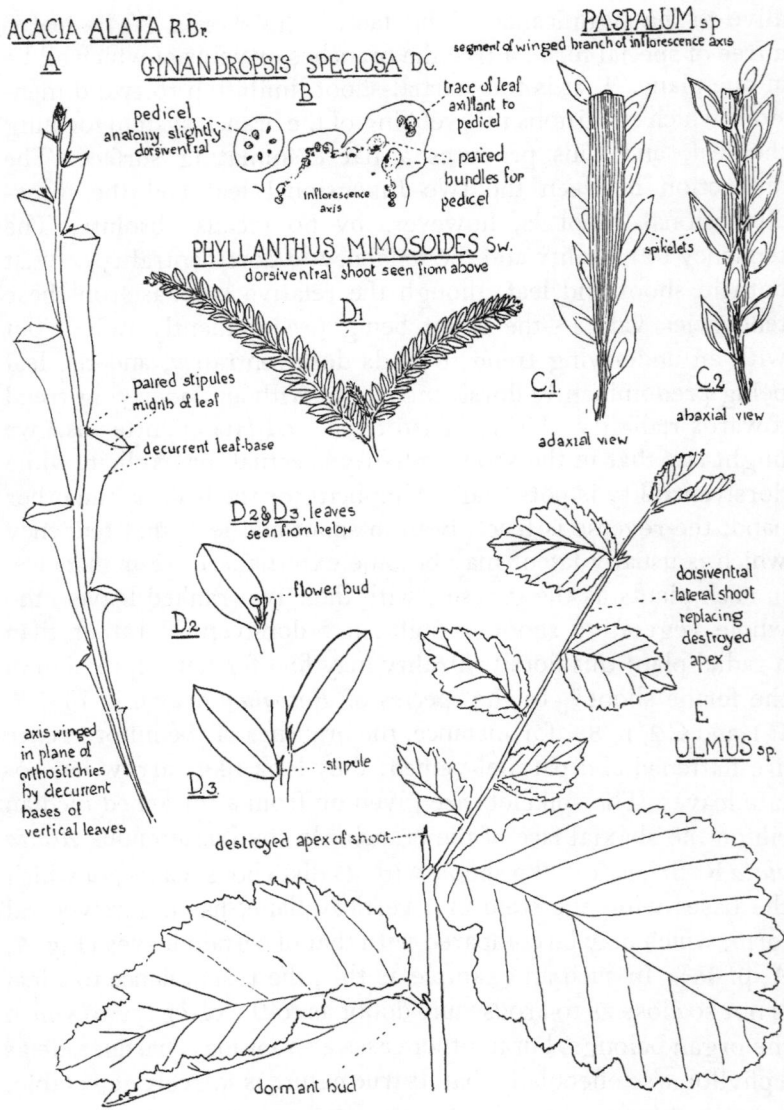

Fig. 7. Dorsiventrality in shoots. A, D1. E ($\times\frac{1}{2}$); B ($\times 14$); C (enlarged, actual width 1·5 mm.); D2, D3 (enlarged).

The morphology of the parts in the bladderworts, for instance, has long been a puzzle to botanists, but the anatomical work of McIntyre and Chrysler[1] upon *Orchyllium* (a terrestrial genus forming part of *Utricularia* sensu lato), suggests that the key to the problem lies in the interpretation of the photosynthetic organs of these plants as shoot structures, though in external features they impersonate leaves with delicate exactitude. Apart, however, from such extreme examples, of which the significance is some-times open to question, we find that ordinary lateral shoots frequently incline to dorsiventrality in their attachment and prophyllar regions. A lateral branch must, indeed, be imperfectly radial at its origin, since a mother-shoot does not normally give rise to a daughter-shoot except from its side, and every lateral shoot thus starts life with an asymmetric relation to its parent. This dorsiventrality of lateral branches is not always confined to the base. In the elm (*Ulmus*),[2] for instance, the primary seedling shoot is radially symmetrical, but all the laterals, on the other hand, are distichous, and have the leaves expanded in one plane (Fig. 7, E, p. 88). In *Coriaria myrtifolia* L. (Fig. 8, B_1, B_2, p. 90), a Mediterranean shrub, the same peculiarity is carried much further. Here the lateral branch systems have an almost deceptive resemblance to compound-pinnate leaves, such as those of *Caesalpinia japonica* Sieb. et Zucc. (Fig. 8, A), but this resemblance is only achieved by degrees in the course of development. In *Coriaria* the leaves in the bud stage are decussate (Fig. 8, B2), but they are brought eventually into one plane by torsion of the internodes and curvature of the leaf-bases. The short branches of some species of *Phyllanthus*, again, simulate compound leaves (Fig. 7, D1, p. 88). Dingler[3] who studied these structures, showed that their marked bilateral development goes with wide divergence from the direction of the parent axis—that is to say, with approximation to horizontality on the part of the daughter shoot. He was also struck by the fact that, among the higher plants in general, shoots of limited growth tend to a flattened form, or at least to a distichous arrangement of their appendages. Leaf-like development of the shoot occurs not only in angiosperms, but

[1] McIntyre, W. G. and Chrysler, M. A. (1943).
[2] Schoute, J. C. (1937).
[3] Dingler, H. (1885); see especially pp. 90, 116.

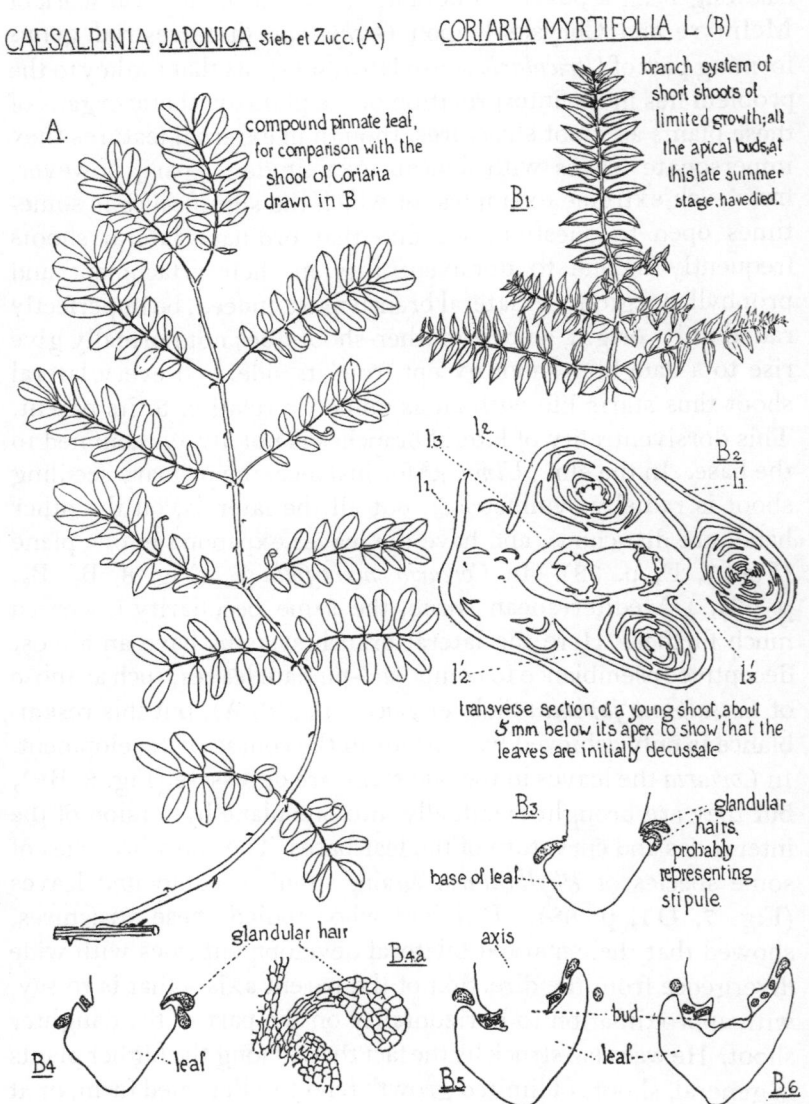

CAESALPINIA JAPONICA Sieb et Zucc.(A)   CORIARIA MYRTIFOLIA L. (B)

A.   compound pinnate leaf,
for comparison with the
shoot of Coriaria
drawn in B

B1.   branch system of
short shoots of
limited growth; all
the apical buds, at
this late summer
stage, have died.

B2. — 1'1.

1₃
1₂
1₁

1'₂

1'₃

transverse section of a young shoot, about
5 mm below its apex to show that the
leaves are initially decussate

B3.   glandular
hairs,
probably
representing
stipule.

base of leaf

glandular hair   axis

B4a.

B4   leaf

bud
leaf

B5.   B6.

Fig. 8. Compound-pinnate leaves and branch systems. A, B1 (× ⅓);
B2 (× 9 *circa*); B3–B6, series through very young leaf-base (× 31 *circa*);
B4a (× 129).

90

in other groups, especially those in which the leaves are small; microphyllous shoots, indeed, lend themselves readily to the simulation of compound leaves. Among gymnosperms, examples are found in *Taxodium* and *Glyptostrobus*,[1] in which certain branches bear 2-ranked leaves without axillary buds; to add to the foliar character of these shoots, they are shed as units in the autumn. The branches, again, of *Thujopsis*, *Thuja*, and some related genera,[1] with their dorsiventral lay-out, and their component phyllomes, shining green above and whitish below, in *facies* recall elaborate leaves. This foliar form of shoot is achieved sometimes in defiance, as it were, of the anatomical orientation. In the pectinate shoot of *Acmopyle*,[2] it has been shown that the flattening of the individual leaves, which gives a phyllome-like appearance to the whole branch, takes place in a plane at right angles to that of leaves in general. The occurrence of leaf-like shoots, associated with microphylly, extends also to the vascular cryptogams. In some species of *Selaginella*,[3] for instance, there is "a frond-like dorsiventral type of branch system", serving for assimilatory purposes, this system being "moulded on lines parallel with those of the fern-frond as a whole".

In the angiosperms, even if a lateral shoot does not show any external dorsiventrality, its anatomy at the base may reveal this tendency. A not uncommon method of vascular connexion between a lateral shoot and its parent, is the division of a shoot bundle into three parts, the median member serving the axillant leaf, while the two wing members of the triad approach one another on the inner face of the leaf bundle, thus forming a bilaterally symmetrical group for the lateral branch.[4] These paired strands have their xylems turned towards one another, so that, through their division, a radial vascular system arises very readily. The leaf, on the other hand, since it is served, as a rule, by an arc or a single bundle, with xylem above and phloem below, is orientated, from the start, for a dorsiventral scheme of anatomy; for, from such an arc or single bundle, simple division

[1] Eichler, A. W. (1889).   [2] Sahni, B. (1920).
[3] Tansley, A. G. (1908).
[4] Bary, A. de (1884), p. 308; and Arber, A. (1932), p. 170, footnote; see also p. 126 of the present book, and Fig. 7, B, p. 88.

of the customary kind will inevitably produce a series of bundles all lying with xylem above and phloem below. The contrast between this type of structure and that of the branch system is not, however, always complete, for it sometimes happens that the paired bundles of the bud-base do not branch in such a way as to radialise the anatomy perfectly; for instance, in the pedicels of the Cruciferae[1] and Capparidaceae, examples may be found of reduction of vascular tissue towards the upper surface, so that the structure shows a hint of dorsiventrality, akin to that of the leaf (cf. *Gynandropsis speciosa* DC., Fig. 7, B, p. 88).

Such considerations as those at which we have been glancing, seem to justify the view that there is a parallelism between leaf and shoot, and that the leaf may be distinguished from the whole-shoot, which bears it, as being a partial-shoot: but the corollary that the leaf, as a partial-shoot, has an urge towards whole-shoot characters, demands a separate discussion, which will be attempted in the succeeding chapter.

[1] Arber, A. (1930), fig. 2, A, p. 307, and (1931 a) fig. 8, A 1, p. 186.

CHAPTER VII

THE URGE TO WHOLE-SHOOT-HOOD IN THE LEAF

THE corollary which we have proposed to the partial-shoot theory—the thesis that the leaf has an urge towards self-completion as a whole shoot—cannot be considered in isolation; we must visualise the leaf and its behaviour on the background of the life of the plant as a whole. The likelihood will be increased that this postulated 'urge' has a real existence, if it can be shown that there is, throughout the shoot-complex and root-complex, a tendency for the members of each generation to compete with, and attempt to dominate, those of the previous generation. The unconscious effort of the leaf to reach the status of the shoot, which is its parent, and even to assume its place, would then no longer be seen as a solitary phenomenon, but as one link in a continuous chain of tendency. Goethe long ago hinted at the general existence of this 'urge'. Some forty years after publishing the *Metamorphose*, he wrote: "When leaves subdivide, or rather when they produce multiplicity out of themselves, this is a striving to become more complete, in the sense that each leaf bethinks itself [gedenkt] to become a branch, as well as each branch to become a tree."[1]

Undoubtedly the *general* rule throughout the plant is that the primary axes dominate their lateral branches. That the growing apex of a shoot exercises a repressive influence upon the lateral buds, is shown by the familiar fact that, if a shoot apex is destroyed, the bud next below it, released from its inhibiting power, often grows out into a replacement shoot; an example from the elm is drawn in Fig. 7, E, p. 88. This usual pre-eminence of the main axis is not, however, the whole story; actually two conflicting tendencies—on the one hand the tendency towards domination by the main axis, and, on the other hand, by the lateral axes—are

---

[1] Troll, W. (1926) [Goethe, J. W. von, *Nacharbeiten und Sammlungen*, pp. 245–6].

93

at work, and the primacy of the main axis is often called in question. In shoot systems, the subordination of parent shoots to lateral shoots is, indeed, quite common; it is the essential feature of all sympodial and cymose branching. As examples we may cite the lime (*Tilia europaea* L.), in which a single lateral shoot each year replaces the terminal bud, which itself dies away, and the lilac (*Syringa vulgaris* L.), in which the terminal bud dies during the summer, and is replaced by two lateral shoots from the axils of the pair of leaves below it. In the case of the lilac, this replacement is the ultimate term in the serial increase in vigour of the lateral buds, which is to be noticed in passing from base to apex of the season's growth. These examples relate to vegetative life, but among fertile shoots we can find corresponding instances, in which a flower (lateral shoot) replaces the apex of the inflorescence (parent shoot). Among the Gramineae, in *Streptochaeta*,[1] *Luziola Spruceana* Benth.,[2] and *Anthoxanthum*,[3] the spikelet axis is actually terminated by a single flower. These cases might be compared with a lime shoot, in which we imagine the replacement of the terminal bud by a lateral to be so complete that the terminal bud is non-existent. On the other hand, in *Lygeum Spartum* Loefl.,[4] the spikelet apex is replaced by two flowers back to back, fitting together so closely that their bracteoles are fused at the base, and no trace of a residual apex belonging to the spikelet is left between them. Such an example might be compared to the lilac shoot, if we suppose the two lateral buds, which take over the work of the terminal bud, to have ousted the latter so completely that they actually unite and fill the space it would have occupied.

The insubordinate behaviour of lateral buds in relation to the shoot apex may extend even into the realm of anatomy; in the inflorescence of certain Fumarioideae, we find that the replacement of the apex of the main shoot by its laterals (the individual flowers) may go so far that the uppermost laterals receive not only their normal supply of bundles, but also any vascular tissue left in the inflorescence axis, so that the apex, if it survives, is non-vascular.[5] Again, in a 20-flowered spike of the curious little

---

[1] Arber, A. (1929a), p. 43.   [2] Arber, A. (1928a), p. 402.
[3] Arber, A. (1927), p. 475.   [4] Arber, A. (1928a), pp. 403–4.
[5] Arber, A. (1931b), p. 340, and fig. 2, D 1–D 5, p. 321.

Juncaginaceous plant, *Lilaea subulata* Humb. et Bonpl., in which the vascular supply was traced in detail,[1] it was found that, whereas the first nineteen flowers received only lateral branches from the main bundles of the inflorescence axis, the whole of the vascular tissue remaining in the axis, after the penultimate flower was supplied, entered the ultimate flower. There was no residual apex to the inflorescence-shoot, and no vascular tissue was left over; the last flower was strictly terminal to the inflorescence, incorporating into itself the whole of the residual vascular tissue of the shoot (3 bundles) instead of receiving merely a single branch bundle.

Innumerable instances of more obvious domination by laterals might be cited from inflorescences, since it is the characteristic feature of all cymose types. As a single example we may mention *Radiola Linoides* Roth (Fig. 9, C, p. 96). The whole plant sketched is little more than an inflorescence, so that it merits its English name, allseed, and its earlier scientific name, *R. Millegrana* Sm. The terminal flower is obscured by successive pairs of laterals, each in turn destined to be overtopped by later branches; six successive generations could be traced in the example drawn. Just the same behaviour may find expression at a higher order of complexity, when an 'inflorescence of inflorescences' is constructed on a cymose plan. This is seen diagrammatically in the composite, *Filago germanica* L. (cudweed), in which Pliny[2] long ago noted that the younger flower heads, like undutiful children, rose above the previous generation, whence, perhaps, it came to be called the impious herb.

In such plants as those which we have cited, the facts are obvious, and their interpretation is not likely to be disputed; but there are other more extreme cases, about which there may be difference of opinion. If, for instance, the main lines of Wydler's[3] explanation are accepted, the shoot system of *Solanum nigrum* L. affords a remarkable example of the early replacement of a parent shoot by a daughter shoot, which is itself rapidly replaced by a grand-daughter shoot, and so on; the whole scheme is complicated

[1] Arber, A. (1940), pp. 618–21, fig. 1, C 1–C 6.
[2] Sillig, J. (1851–8), vol. IV, 1855, lib. 24, cap. 19, sect. 113, p. 99; Holland, P. (1601), p. 205.
[3] Wydler, H. (1844) and (1857).

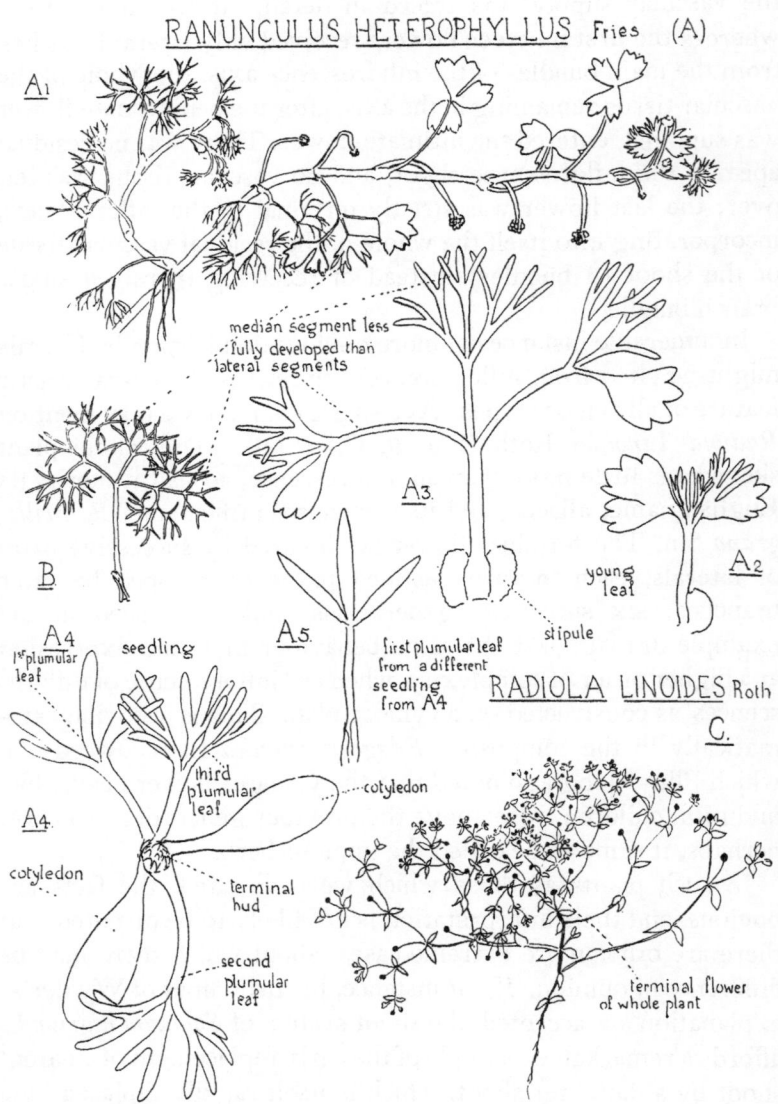

RANUNCULUS HETEROPHYLLUS Fries (A)

A1

median segment less
fully developed than
lateral segments

A3.

B

young
leaf

A2

A4
1st plumular
leaf

seedling

A5.

first plumular leaf
from a different
seedling
from A4

stipule

RADIOLA LINOIDES Roth

C.

third
plumular
leaf

cotyledon

A4.

cotyledon

terminal
bud

second
plumular
leaf

terminal flower
of whole plant

Fig. 9. Predominance of laterals over parent axes in leaves and shoots.
A1 ($\times\frac{1}{2}$); A2, A3 ($\times 2.4$ circa); A4, A5 (enlarged); B, another species
of Ranunculus (nat. size); C, semi-diagrammatic (slightly reduced).

by shoot and leaf fusions. On Wydler's theory, the succession of lateral axes, each generally bearing two prophyllar leaves only, forms itself into a structure deceptively like a main axis of continuous growth.

Turning from the shoot to the leaf, we find that, just as the lateral shoot may dominate, or even incorporate the apex of the parent shoot, so a leaf may overpower, or even absorb the apex of the shoot of which it is a part. The sterile phylloclade of certain Liliaceae—*Ruscus* (butchers' broom), and the related *Danae* and *Semele*[1]—can be interpreted as a short shoot, the apex of which is embodied in its own solitary leaf (prophyll). Such a phylloclade may terminate a long shoot, and it is then an example both of a lateral shoot replacing the apex of its parent shoot, and of a prophyll overmastering the secondary shoot of which it is the only leaf member. Probably the needle of *Asparagus* comes into the same category as the phylloclade of *Danae*, *Ruscus*, and *Semele*; the needle would then be regarded as a short shoot reduced to its prophyll, which in this genus is petiole-like.[2]

The attempt of the leaf to dominate the parent-shoot is revealed even in early ontogeny. It has been shown that some influence from the leaves has an inhibiting effect upon both axillary and terminal buds, causing them, after a time, to stop elongating, and turn into winter-buds.[3]

The subordination of the shoot apex to its own lateral shoots, or to its own leaves, may be paralleled within the leaf itself, in the domination of the leaf-apex by lateral leaflets or lobes. The median region of a simple leaf, or the terminal pinna or rachis-tip of a compound leaf, may be considered as the primary part of the leaf, since it is served directly by the midrib, whereas the rest of the leaf is served by lateral branches from the midrib. In the different forms of leaf met with in *Ranunculus heterophyllus* Fries (Fig. 9, A1–A5, p. 96), the apical lobe, despite its primacy, tends to be reduced; it is sometimes filiform, even when the lateral lobes show an expanded lamina. Correspondingly, in some compound-pinnate leaves, the end pinna may be reduced or absent.

---

[1] Arber, A. (1924 *a*).

[2] The interpretation offered in Arber, A. (1924 *b*) and (1935) is here corrected after a reconsideration of Stefanoff, B. (1932).

[3] Snow, R. (1929) and references there given.

In *Bauhinia yunnanensis* Franch. (Fig. 10, A 1, A 2, below), and *Passiflora capsularis* L. var. *acutiloba* (Fig. 10, C), there are two lateral pinnae or lobes, alone, the terminal pinna or lobe being reduced to a mere point, while, in *Gleditschia caspica* Desf., both terminal pinna and pinnule are abortive (Frontispiece, B 4); moreover, in *Amicia Zygomeris* DC. (Fig. 6, A 1, p. 81), the

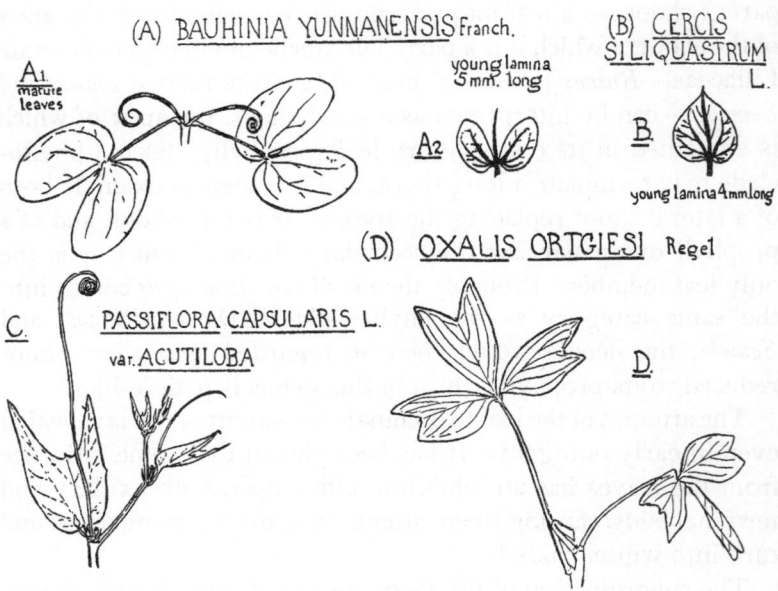

Fig. 10. A, *Bauhinia yunnanensis* Franch.: A 1, pair of mature leaves, apical points dominated by lateral pinnae (× ½); A 2, young lamina (enlarged) to show relative importance of the apical point in the early stages. B, *Cercis Siliquastrum* L., young leaf (enlarged) for comparison with A 2. C, *Passiflora capsularis* L. var. *acutiloba*, end of shoot showing leaves with apical points dominated by lateral lobes (× ½). D, *Oxalis Ortgiesi* Regel, end of shoot showing apical points of pinnae, dominated by lateral lobes (× ½).

terminal pinna is entirely absent. The process of inhibition of the primary apical region of the phyllome may even go so far that certain apparently simple leaves may be interpreted as consisting of two lateral pinnae in a state of union, the median pinna being absent. The foliage of the Judas tree (*Cercis Siliquastrum* L.), for instance, has been explained on these lines. A very young leaf of this species is shown in Fig. 10, B, above; the halves (divided by the median line of the midrib) are held to correspond to free lateral lobes, such as those of *Bauhinia yunnanensis* Franch. (Fig. 10, A 1 and A 2).[1]

Sometimes it is not lateral pinnae, but stipules, which gain ascendancy over the rest of the leaf. In *Lathyrus Aphaca* L.,[2] the seedling leaves bear pinnae (e.g. leaves 3 and 4, Fig. 11, A 4, p. 100), but, in the mature stages, a rachis-tendril (pinnaless as a rule) takes the place of the leaflet-bearing rachis (Fig. 11, C), so that the stipules alone represent the laminate part of the phyllome. In plants belonging to other families the lamina may be subordinated even to vanishing point. A stage towards complete disappearance is seen in the shoot of *Polygonum equisetiforme* Sibth. et Sm., in which, at some nodes, the phyllome is reduced to the leaf-base with its chaffy stipular ochrea, no blade being detectable at maturity. More extreme examples occur among the Cruciferae. The watercress (*Nasturtium officinale* R.Br.) is generally described as having a bractless raceme, but actually, though the lamina of the bract, which should be associated with each flower, is lacking, microtome sections show that an ephemeral squamule arises from the axis on either side of the base of each pedicel. These can scarcely be interpreted except as the stipules of the absent bract, so that we have here a case of leaf reduction carried to such a point that the median region of the phyllome has vanished entirely, and it is only the paired basal laterals—the stipules—which survive in a non-vascular condition.[3] Moreover stipular structures may not only remain the sole representatives of the phyllomes of which they form a part— they may even mimic, as it were, their whole-leaf characters. For instance the ligular sheath of the water hyacinth (*Eichhornia speciosa* Kth.) has a terminal lobe suggesting a miniature lamina.[4]

Not only may there be reduction of the median, in relation to the lateral regions, in the leaf as a whole: a comparable reduction may also occur in the individual members of a compound

[1] Velenovský, J. (1905–13), vol. II, 1907, p. 497; cf. also Fries, R. E. (1909).

[2] There is a full and interesting account of the morphology of this plant in Barnes, B. (1933).

[3] For a study of stipules in the Cruciferae, see Norman, J. M. (1857), (1858), and Arber, A. (1931a), fig. 8, p. 186; in the legend of this figure, lines 4 and 5, and p. 187, line 7, 'leaves' should be corrected to 'pedicels'. See also, in the present book, p. 141 and fig. 26, E, p. 140. In 1931 Norman's work was unknown to the present writer.

[4] Arber, A. (1922a), pl. I, fig. 8; or (1925), fig. lxxxi, 8, p. 108.

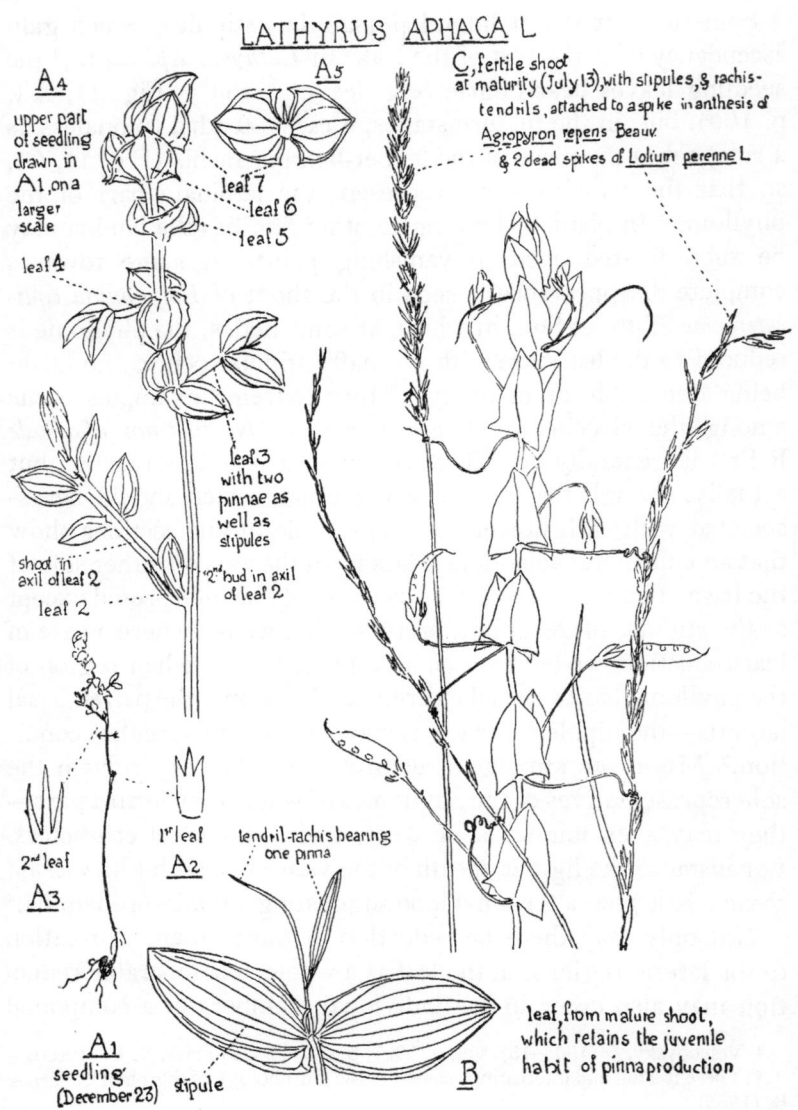

LATHYRUS APHACA L

A4
upper part
of seedling
drawn in
A1, on a
larger
scale

A5

leaf 7
leaf 6
leaf 5

leaf 4

leaf 3
with two
pinnae as
well as
stipules

shoot in
axil of leaf 2
leaf 2

2ⁿᵈ bud in axil
of leaf 2

C, fertile shoot
at maturity (July 13), with stipules, & rachis-
tendrils, attached to a spike in anthesis of
Agropyron repens Beauv.
& 2 dead spikes of Lolium perenne L

2ⁿᵈ leaf

1ˢᵗ leaf
A2

A3

tendril-rachis bearing
one pinna

A1
seedling
(December 23)   stipule

B

leaf from mature shoot,
which retains the juvenile
habit of pinna production

Fig. 11. *Lathyrus Aphaca* L., dominance of stipules. A 1, C (×½);
A 2–A 5 (enlarged); B slightly enlarged.

100

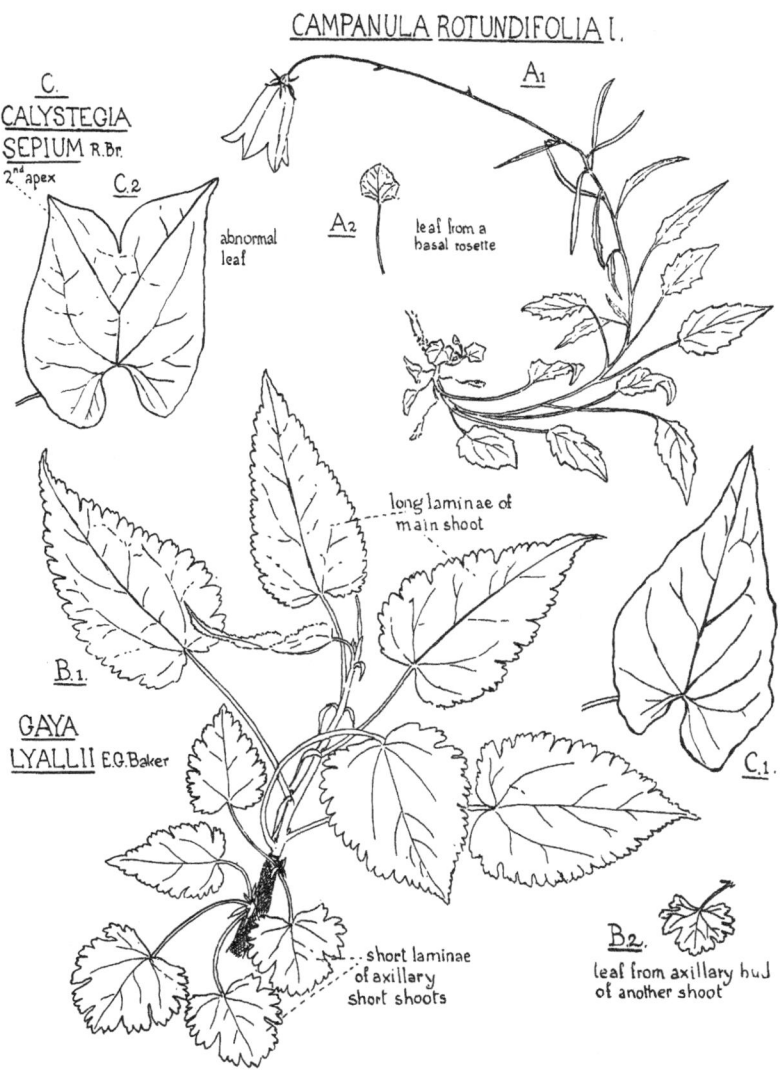

CAMPANULA ROTUNDIFOLIA L.

C.
CALYSTEGIA
SEPIUM R.Br.
2nd apex
C.2

A1

A2

abnormal
leaf

leaf from a
basal rosette

long laminae of
main shoot

B.1.

GAYA
LYALLII E.G.Baker

C.1.

short laminae
of axillary
short shoots

B2.
leaf from axillary bud
of another shoot

Fig. 12. Relative importance of median and lateral veins.
All drawings ($\times \frac{1}{2}$).

101

leaf. In *Oxalis Ortgiesi* Regel (Fig. 10, D, p. 99), for instance, the pinnae of the ternate leaf recall in their form the whole leaf of *Passiflora capsularis* L. (Fig. 10, C).

When we study the nervation of the leaf, rather than its outline, we again find the same factor coming into play. As we have already noticed, it has been shown[1] that there is an actual correlation between the subordination of the main shoot to lateral shoots, and the subordination of the leaf midrib to lateral veins; that is to say, when the shoot branching is cymose, there is a tendency for the leaves to be palmately nerved. Palmate venation may be envisaged as a type in which the main vein is shortened by the suppression of its earlier 'internodes'; the laterals are thus more important relatively to the midrib than in pinnate leaves, in which the midrib is elongated and dominates its own branches. The supremacy of the midrib may vary in degree even within a single plant. Fig. 12, A and B, p. 101, shows the range from palmate to more or less pinnately veined leaves in *Campanula rotundifolia* L. and *Gaya Lyallii* E. G. Baker. In both these species the pinnate form seems to be the more fully developed. A corresponding change may, moreover, be witnessed within an individual phyllome. In *Agrimonia Eupatoria* L. the compound leaf shows an alternation[2] of relatively large pinnately veined pinnae, and small palmately veined pinnae, in which the midrib does not exceed the laterals (Fig. 13, D 1 and D 2, p. 103). In certain peltate leaves, reduction of the midrib can again be observed. *Diphylleia cymosa* Michx. is an example (Fig. 13, A 1, p. 103); its condition may be likened to that of the petal or honey-leaf of *Berberis calliantha* Mull., belonging to the same family (Fig. 13, C). Here the median vein is small in comparison with the vascular strands which supply the massive lateral nectaries. A further stage is reached in *Podophyllum peltatum* L., where phyllomes may occur in which the midrib is absent (Fig. 13, B). Other examples of laterals exceeding the midrib in importance are found in the anatomy of *Acacia* phyllodes and various Irid leaves in which the main laterals jointly form a pseudo-midrib.[3]

---

[1] Uittien, H. (1928a); cf. also p. 83.   [2] See p. 123.

[3] Arber, A. (1921), *p.m.r.* in figs. 40–6, p. 318; or (1925), fig. lxxix, 40–6, p. 105. For examples of reduced midribs see (1921), figs. 36–8, p. 315; or (1925), fig. liv, 36–8, p. 76.

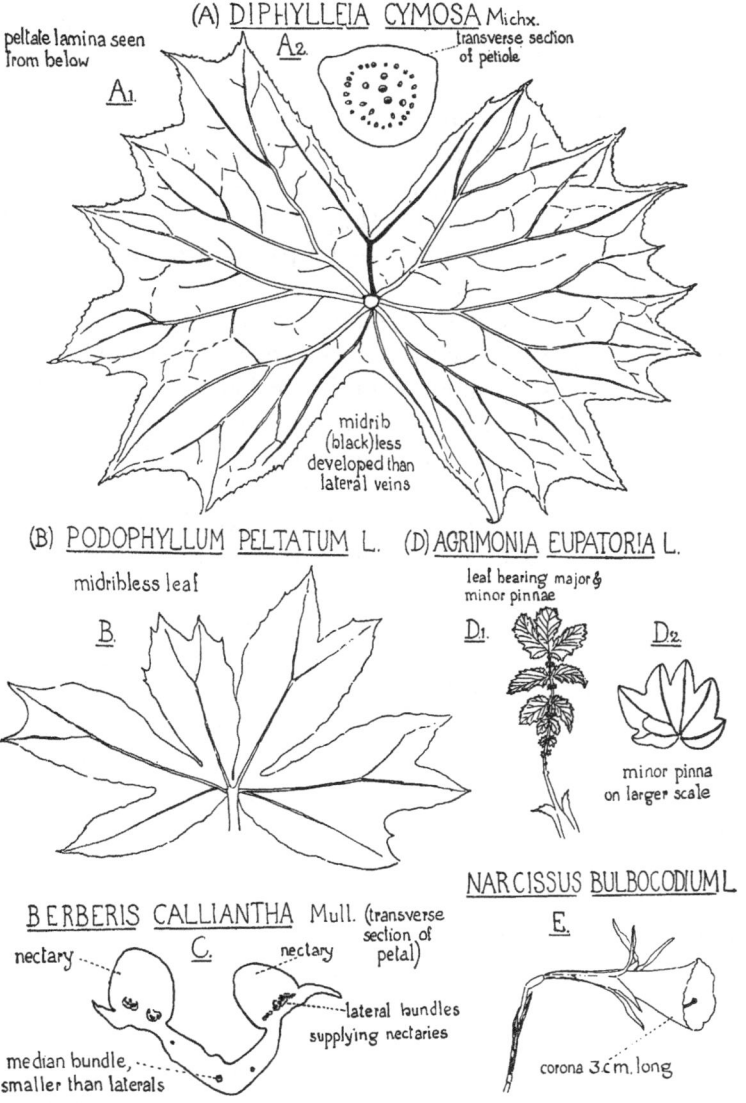

(A) DIPHYLLEIA CYMOSA Michx.

peltate lamina seen
from below

A₁.

A₂.

transverse section
of petiole

midrib
(black)less
developed than
lateral veins

(B) PODOPHYLLUM PELTATUM L.

midribless leaf

B.

(D) AGRIMONIA EUPATORIA L.

leaf bearing major &
minor pinnae

D₁.

D₂.

minor pinna
on larger scale

BERBERIS CALLIANTHA Mull. (transverse
section of
petal)

nectary

C.

nectary

lateral bundles
supplying nectaries

median bundle,
smaller than laterals

NARCISSUS BULBOCODIUM L

E.

corona 3 cm. long

Fig. 13. A 1, B, D 1, E (× ½); A 2 (× 2 *circa*); B, one of a pair of phyllomes
below a flower; C (× 15 *circa*); D 2 (enlarged).

103

Occasionally, as an abnormality, the supremacy of the midrib, in a leaf in which it is usually dominant, may be contested by one of the laterals, which forms, as it were, a second leaf apex, side by side with the first. An example from the convolvulus is drawn in Fig. 12, C 1 and C 2, p. 101.

If the relation of an enation to its parent leaf is accepted as being of the same nature as the relation of the leaf itself to its parent shoot, we may regard such an exaggerated corona[1] as that of *Narcissus Bulbocodium* L. (Fig. 13, E, p. 103) as an example in which a partial-leaf dominates its parent whole-leaf, just as a whole-leaf may dominate its parent shoot.

Reproductive phyllomes—petals, stamens, and carpels—are frequently ternate in anatomical structure, while, in their external form, they show subordination of the median region much more markedly than foliage-leaves. This phenomenon is perhaps an expression of the tendency to inhibition of growth in length, which is one of the essential characters of the flower-shoot itself, and sometimes reappears in the inflorescence-shoot (cf. Compositae). Many petals tend to be heart-shaped, with the distal marginal region overgrown in comparison with the median part. The paired pollen-sacs of the stamen may again be regarded as lateral developments overriding the main apex, while it is the lateral wings of the carpels—rather than their median regions—on which falls the whole onus of producing the ovules. As would be expected from this general scheme of carpel organisation, it often happens that the lateral bundles are much more conspicuous than the midrib. The poppy (*Papaver*) is a striking instance. Here the median region of the carpel is so much reduced that it is the margins alone which reach the apex and form the stigmas, while the median bundle is poorly developed in comparison with the main laterals.[2]

In the detailed structure of the vascular system itself, we also meet with features suggesting the urge of lateral members to attain the status of their parents. The replacement of a dorsiventral collateral bundle by a radial concentric strand, or a bundle ring, may be interpreted as a striving after the stelar condition—i.e. after the anatomical characteristics of a whole-shoot—on the part of individual strands, which are themselves merely sub-

[1] Cf. Arber, A. (1937 *b*).   [2] Arber, A. (1938 *b*), fig. 1, p. 651.

sidiary members of a vascular system. The Leguminosae offer examples of this (cf. *Amherstia*, Fig. 14, A 1 and A 2, p. 106).

We have now followed in several fields the tendency of lateral structures to aspire to take up the role of the parent, for which they are, as it were, understudies. We have seen this tendency in the relation of a vegetative or reproductive shoot of one order to the shoot of the previous order; in the relation of a leaf to the shoot which bears it; in the relation of branches (pinnae) of the leaf to the main rachis with its terminal pinna; in the relation of both stipules and lateral lobes of a simple leaf to the median region; in the relation of the lateral lobes of the pinna of a compound leaf to the median part of the pinna; in the relation of the lateral veins to the median vein in vegetative and reproductive phyllomes; in the relation of a partial-leaf (enation) to its parent whole-leaf; and in the relation of individual bundles to a complete vascular system. It is thus evident that the urge to whole-shoot characters, which we have postulated for the leaf, is only one instance of the tendency of lateral structures to assume the characters of the parent—a trend which can be detected in a shoot in all its phases and components. Having shown our grounds for regarding this as a valid general principle, we have next to turn to the special evidence for the existence of this particular urge in the leaf, when it is considered in relation to the shoot.

If we set aside, for the moment, the new theory of the axillary bud proposed in the succeeding chapter (pp. 125–131), and adopt the existing view that the lateral bud is derived from the stem rather than the leaf, we may say that the majority of leaves show their reduced character, as compared with shoots, in their inability to give rise either to partial or to whole-shoots. There are, however, many exceptions, in which the urge to vegetative reproduction manifests itself in phyllomes. A cabbage leaf, for instance, may produce shoot appendages; a foliar cluster derived from the distal region of the leaf is drawn in Fig. 14, B, p. 106, and a leaf-bearing outgrowth from the midrib in Fig. 15, B 4 and B 5, p. 110. Even more striking cases of shoot-bearing cabbage leaves have been described; in one of these, as many as five leafy branches sprang from the midrib.[1] In the tomato (*Lycopersicum esculentum* Mill.) similar developments have been recorded.[2]

[1] Duchartre, P. (1881), p. 257. [2] Penzig, O. (1921, 1922), vol. III, p. 70.

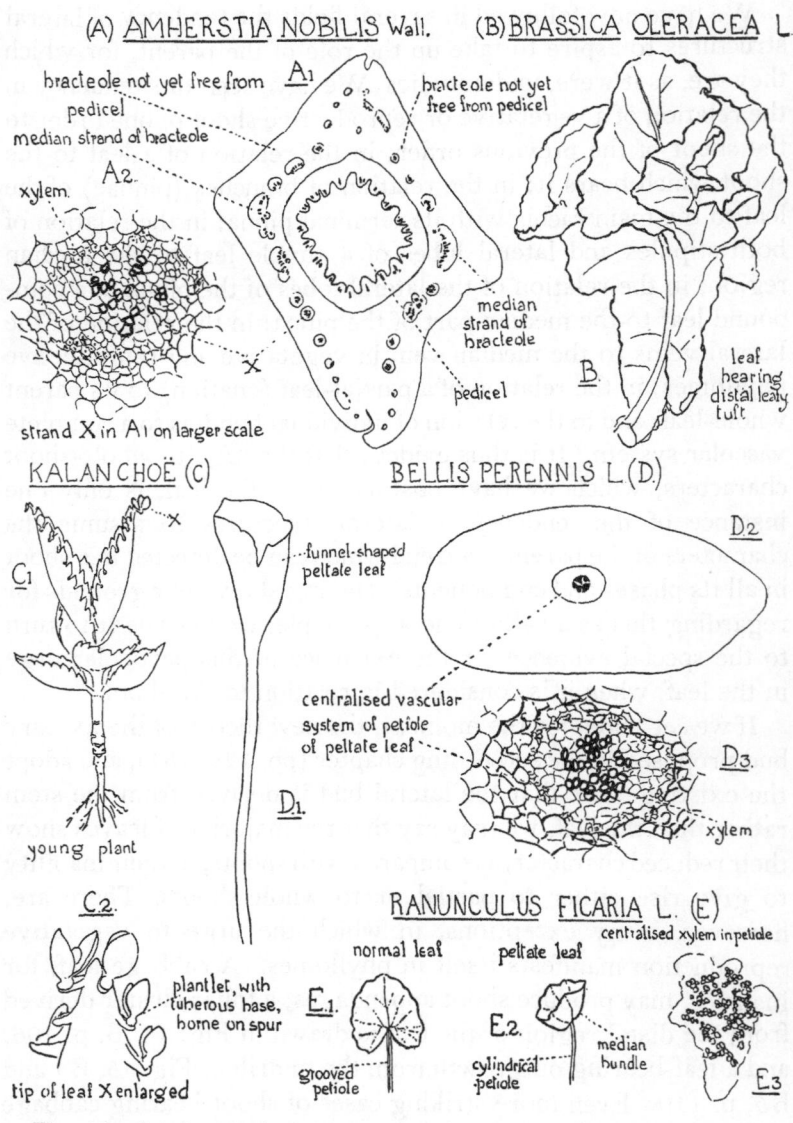

(A) <u>AMHERSTIA NOBILIS</u> Wall.   (B) <u>BRASSICA OLERACEA</u> L.

bracteole not yet free from pedicel — A_1

median strand of bracteole

A_2

xylem

strand X in A1 on larger scale

bracteole not yet free from pedicel

median strand of bracteole

pedicel

B.

leaf bearing distal leafy tuft

<u>KALANCHOË</u> (C)

C_1

young plant

C_2

plantlet, with tuberous base, borne on spur

tip of leaf X enlarged

<u>BELLIS PERENNIS</u> L.(D)

funnel-shaped peltate leaf

centralised vascular system of petiole of peltate leaf

D_1.

D_2

D_3.

xylem

<u>RANUNCULUS FICARIA</u> L. (E)

centralised xylem in petiole

normal leaf   peltate leaf

E_1

E_2

grooved petiole   cylindrical petiole

median bundle

E_3

Fig. 14. A1 ($\times 14$ *circa*); A2 ($\times 193$ *circa*); B, C1, E1, E2 ($\times\frac{1}{2}$); C2 (enlarged); D1 ($\times 3\frac{1}{2}$ *circa*); D2, E3 ($\times 47$); D3 ($\times 193$ *circa*).

Indeed, in leaves in general, it is probable that the power of shoot production is latent rather than absent; leaf-cuttings of many plants can be induced, by appropriate cultural methods, to root and give rise to shoots.[1] There are also a number of examples of leaf-budding occurring as a regular event. Remarkably active plantlet formation happens at the marginal notches of the leaf in *Kalanchoë daigremontiana* R. Hamet et Perrier de la Bâthie (Fig. 14, C 1 and C 2, p. 106); a large leaf may bear as many as 60 offspring. So strong is the reproductive urge that these buds, while still attached to the parent leaf, may themselves produce plantlets of the second order on their own small leaves.[2] Aquatics are particularly liable to leaf-budding. Fig. 18, F 1, p. 115, illustrates shoot production from the leaf of *Nymphaea stellata* Willd.; this occurs immediately above the insertion of the lamina, and thus at the fountain head of the vascular supply (Fig. 18, F 2). It is noticeable that, in all these cases of leaf-budding, it is a new shoot, and not merely a new leaf, which is produced; this may be regarded as an indication of latent whole-shoot character in the leaf.

Even in its embryonic phases, the leaf shows an urge towards whole-shoot characters, for its apical growth in length, in early stages, may be interpreted as an effort after the continued elongation of a shoot apex, though this effort soon dies down.

Another way in which the leaf may attempt, as it were, to rival a whole-shoot, is by an approach to radiality in its mature form. The usual type of dorsiventral leaf develops a single lamina-wing to right and left, so that its expansion is all in one plane. Examples, however, sometimes occur in which extra wings are produced. The blade of *Gladiolus tristis* L.,[3] for instance, is cruciate in section, and there are other examples among mono-cotyledons of complex leaf types, which suggest the radial character of the shoot rather than the flattened character of the leaf. Moreover some leaves, which are usually dorsiventral, may occasionally show a hint of radialisation. The cabbage, for instance, not infrequently develops minor lamina wings from

[1] See, for instance, Graham, R. J. D. and Stewart, L. B. (1929).
[2] Johnson, M. A. (1934).
[3] Arber, A. (1921), fig. 48, J, p. 318; or (1925), fig. lxxix, 48, J, p. 105.

two lateral lines on the upper surface of its substantial midrib, in addition to the wings forming the normal blade.

The tendency of the leaf to assume the status of the parent shoot is revealed, not only in its behaviour and in its external form, but also in its anatomical construction. When we examine the dorsiventral petiole of a typical leaf, with its arc of bundles resembling in section a segment of a stem ring, we see that it has all the appearance of an incomplete shoot. Many petioles, however, are not simply dorsiventral, but show a pronounced tendency towards a radiality suggesting a whole-shoot; for, even if the bundles form an arc at the base, at a higher level this may be replaced by a ring. This replacement may be brought about in various different ways, as though the petioles had a radial structure as their common goal, but reached this goal by a variety of routes.

Among the monocotyledons, many complete leaves are more or less cylindrical, and in their anatomy show the same radial trend as the petioles of dicotyledons, a fact which formerly led the writer to describe them as petiolar phyllodes;[1] but, in the light of more recent knowledge, it seems better to modify this interpretation, and to regard such a leaf as *a fixation of the whole phyllome at its pre-laminar stage*. In this early phase the primordium has been described as often taking the form of a "tapering, adaxially-flattened cone",[2] thus recalling a petiole, from which however it differs in so far as it is potentially a whole phyllome, whereas the petiole is merely one element in this ultimate whole. The idea of such fixation at immature stages may have its relevance in the comparative interpretation of plant forms; Takhtajan[3] has, for instance, suggested that the herbaceous habit may be treated as a persistent juvenile phase of the tree habit.

In so-called 'unifacial' petioles, and in those monocotyledonous leaves which show the same character in what would ordinarily be the laminar region, the upper face of the phyllome is entirely eliminated throughout most of its length, since the lateral junction lines of the upper and lower surfaces close around the phyllome at the top of the leaf-base, and actually meet in the upper median line; a condition approximating to this is seen in Fig. 18, D2,

---

[1] Arber, A. (1918), etc.; and (1925), p. 100. On what was formerly called a pseudo-lamina, see p. 85 of the present book.
[2] Foster, A. S. (1936*a*).   [3] Takhtajan, A. (1943).

p. 115. The whole external surface in unifacial petioles and leaves thus corresponds to the external surface of the parent shoot, and, if there is also a closed ring of bundles, the structure approaches nearly to the radial character of the stem. The leaf of *Tropaeolum majus* L. is, in a certain respect, an extreme example, for here the leaf-base shares with the petiole an entirely unifacial structure.[1]

It was pointed out by Casimir de Candolle[2] that a certain type of lamina—the shield-shaped or peltate—tends to be correlated with a unifacial petiole, which has "un système fibrovasculaire complet, c'est-à-dire également réparti tout autour de l'axe de figure". Such a vascular scheme is an approximation to stem structure. Instances of more or less completely radial anatomy, associated with various kinds of peltation, may be seen in Fig. 13, A2, p. 103; Fig. 16, A5 and B3, p. 112; and Fig. 17, C3, p. 113. In the not infrequent examples in which a plant, which normally has non-peltate laminae and bifacial petioles, produces a peltate or ascidial lamina as an abnormality, its petiole will generally be found to have diverged from the form typical for the species, and to have become unifacial. The drawings in Fig. 14, D1, E1, E2, p. 106, show how abnormal peltate leaves, of the cup- or funnel-shaped types, differ from normal leaves in the daisy (*Bellis perennis* L.) and the lesser celandine (*Ranunculus Ficaria* L.). In both cases the bundles in the normal petiole are arranged in an arc, while, in the peltate or ascidial form, the whole of the xylem is grouped into a central core, and the structure is more or less radial (Fig. 14, D2, D3, E3).

In the abnormal leaves of *Platanus acerifolia* Willd. and *Brassica oleracea* L., sketched in Fig. 15, p. 110 (A1, A2, B1–B3) the midrib detaches itself from the lamina, and after being free for a certain distance, produces an ascidial peltate blade. At first glance, such an appearance may suggest the inflorescence of the lime[3] (*Tilia*, Fig. 38, p. 176), in which the axis is fused for part of its length with its own bracteole; but we have here a failure in disjunction between a shoot-axis and a phyllome midrib, followed by their separation at a higher level, whereas these abnormal

[1] Troll, W. (1932), p. 167.
[2] Candolle, A. Casimir P. de (1899a); see also (1868) and (1899b).
[3] Arber, A. (1924a), p. 256, and fig. 49, p. 257.

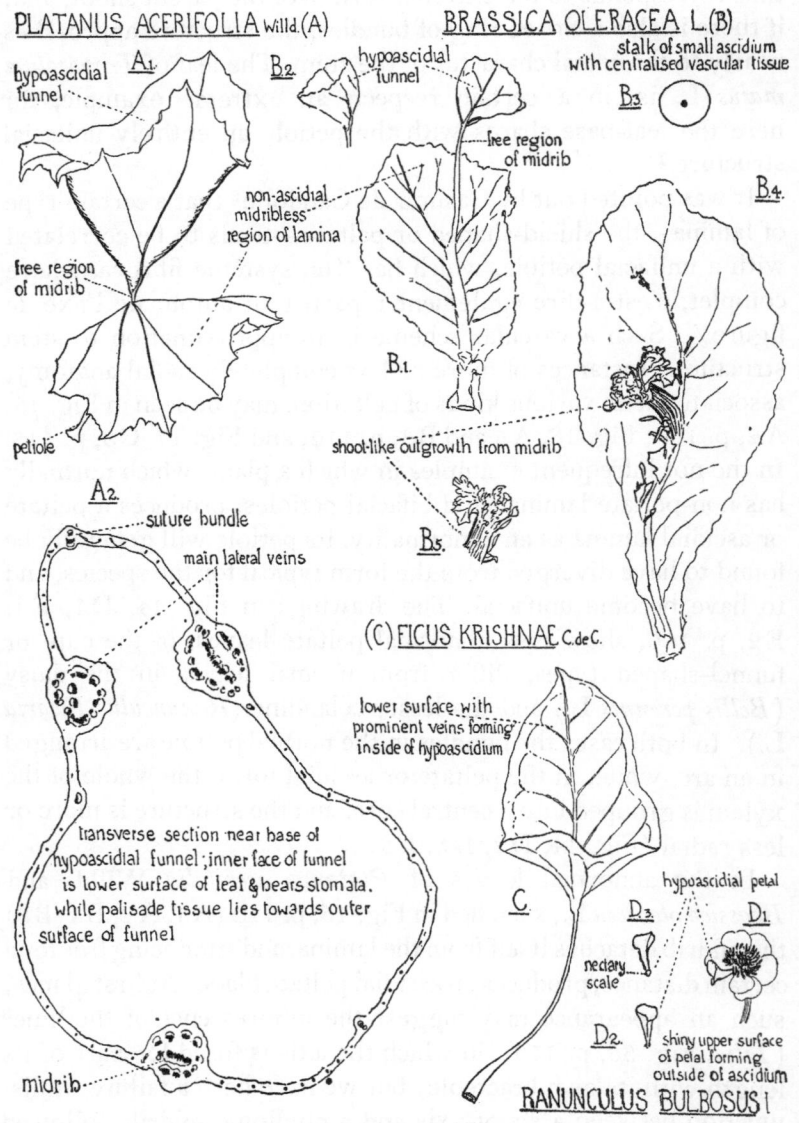

PLATANUS ACERIFOLIA Willd.(A)

hypoascidial funnel

A₁.

B₂.

free region of midrib

petiole

A₂.

suture bundle

main lateral veins

transverse section near base of hypoascidial funnel; inner face of funnel is lower surface of leaf & bears stomata, while palisade tissue lies towards outer surface of funnel

midrib

non-ascidial midribless region of lamina

BRASSICA OLERACEA L.(B)

hypoascidial funnel

stalk of small ascidium with centralised vascular tissue

B₃.

free region of midrib

B₄.

B₁.

shoot-like outgrowth from midrib

B₅.

(C) FICUS KRISHNAE C.deC.

lower surface with prominent veins forming inside of hypoascidium

C

D₃.

nectary scale

D₂.

hypoascidial petal

D₁.

shiny upper surface of petal forming outside of ascidium

RANUNCULUS BULBOSUS l

Fig. 15. A 1, B 1, B 2, B 4, B 5, D (× ½); A 2 (× 7 circa); B 3 (× 9 circa); C (× ⅓), hort. var. of *F. benghalensis* L., see Biswas, K. (1935).

plane and cabbage leaves represent a genuinely shoot-like development of the midrib itself, in which no other member is concerned. Such abnormalities, as well as unifaciality in petioles and peltation in normal leaves, and in carpels, may be regarded as expressions of an urge towards the radialness of the whole-shoot; Casimir de Candolle spoke of the peltate as "le type des phyllomes les plus développés".[1] A compound leaf with a peltate lamina often, indeed, looks strikingly like a shoot-axis bearing a distal tuft of leaves; this comparison was long ago suggested for the compound-pinnate leaf of the lupin[2] (Fig. 17, A 1 and A 2, p. 113). The build of this leaf approximates closely to that of a pseudo-whorl of *Asperula odorata* L. (Fig. 17, B), below an unexpanded terminal bud. Certain species of *Oxalis*[3] show a structure corresponding to that of the lupin—one of these species has indeed been named *O. lupinifolia* Jacq. In *O. ennea-philla* Cav. (Fig. 16, B 1–B 3, p. 112) there may be a dozen leaflets crowded round the petiolar apex in such a way that it is difficult, in a detached leaf, to determine which leaflets are abaxial and which adaxial. *Akebia quinata* Decne. (Fig. 17, C 1–C 4, p. 113) and *Tropaeolum pentaphyllum* Lam. (Fig. 16, A 1–A 6) furnish other examples of compound leaves approaching or reaching full peltation. Fig. 16, A 1 shows that the lamina of *T. penta-phyllum* sometimes has five leaflets, so that there is a front *pair*, and the structure is thus only incompletely radial. On the other hand, the two front leaflets may be fused partially (Fig. 16, A 2), or wholly (Fig. 16, A 3), thus bringing the radiality nearer to completion. An analogy which, fantastic as it sounds, may yet have a certain suggestive value, might be traced between certain peltate leaves and reproductive shoots. Those species of *Geranium* and *Pelargonium*, for instance, which have two stipules, and a petiole surrounded at the apex by a peltate lamina, may perhaps be compared with a reproductive shoot with two bracteoles, a pedicel, and a calyx surrounding the apex of the pedicel. Beyond the calyx no further comparison can be drawn, since the axis, in the case of the flower, continues to grow and produces the remaining floral members, whereas lamina-production

[1] Candolle, A. Casimir P. de (1899 a).

[2] Schultz-Schultzenstein, K. H. von (1861), p. 281.

[3] On *Oxalis lasiandra* Zucc. see Troll, W. (1935 etc.), vol. I, pp. 1794–5.

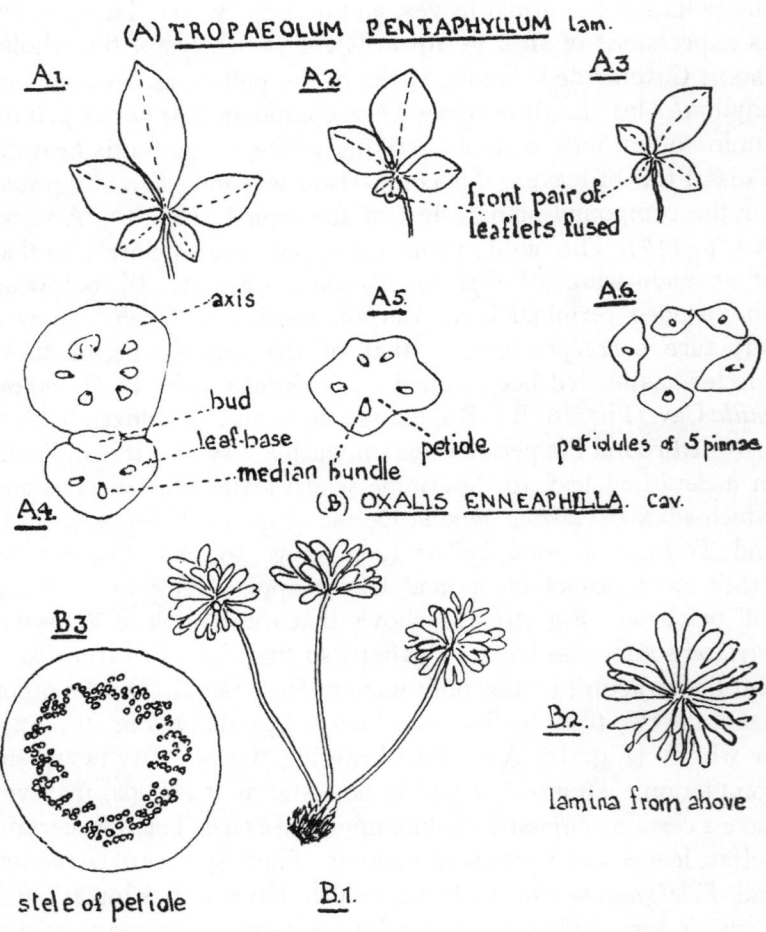

(A) TROPAEOLUM PENTAPHYLLUM lam.

A1.

A2.

A3.

front pair of leaflets fused

axis

A5.

A6.

bud

leaf-base

petiole

petiolules of 5 pinnae

median bundle

A4.

(B) OXALIS ENNEAPHILLA. Cav.

B3

B2.

lamina from above

stele of petiole

B.1.

Fig. 16. A1–A3, slightly enlarged; A4–A6 (× 46); B1, B2 (nat. size); B3 (× 28), the outline is that of the central cylinder, and lignified elements alone are indicated.

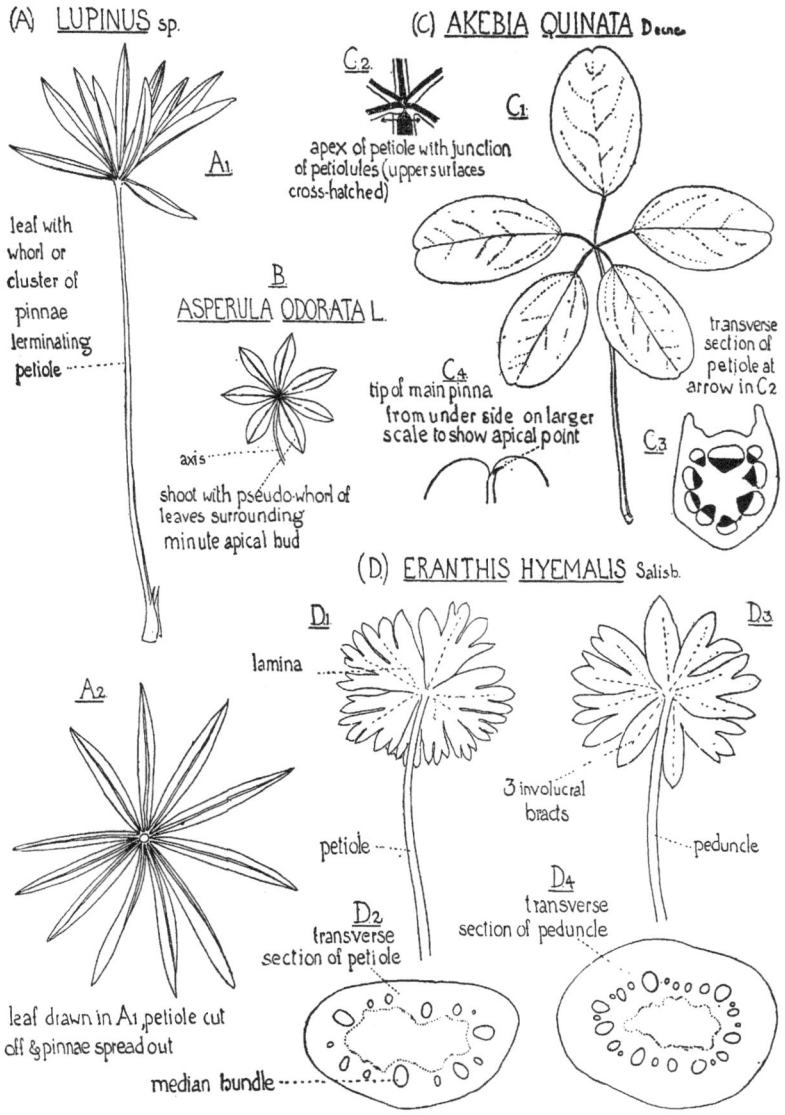

(A) LUPINUS sp.

A₁

leaf with
whorl or
cluster of
pinnae
terminating
petiole

A₂

leaf drawn in A1, petiole cut
off & pinnae spread out

B.
ASPERULA ODORATA L.

axis

shoot with pseudo-whorl of
leaves surrounding
minute apical bud

(C) AKEBIA QUINATA Deme.

C₂.

apex of petiole with junction
of petiolules (upper surfaces
cross-hatched)

C₁.

C₄.
tip of main pinna
from under side on larger
scale to show apical point

transverse
section of
petiole at
arrow in C2

C₃.

(D) ERANTHIS HYEMALIS Salisb.

D₁

lamina

D₃

3 involucral
bracts

petiole

peduncle

D₂
transverse
section of petiole

D₄
transverse
section of peduncle

median bundle

Fig. 17. A 1, A 2, B, C 1, D 1, D 3 (× ⅓); C 2, C 4 (on larger scale than C 1);
C 3, D 2, D 4 (× 9 *circa*).

exhausts the resources of the phyllome. Certain other leaves, which approach peltation without actually achieving it, are also strikingly shoot-like. The foliage-leaf of *Eranthis hyemalis* Salisb. (Fig. 17, D1, D2, p. 113), for instance, with its long petiole, radial in anatomy, and its spreading and subdivided blade, recalls the peduncle of the same plant, with its crown of three associated bracts (D3, D4); to complete the resemblance, one must suppose the flower to be undeveloped.

The notion that there could be any morphological significance in similarities, such as those we have been citing, between members which are not homologous in the phylogenetic sense, would have seemed ridiculous in the nineteenth century, during the dominance of the Darwinian theory. These parallels exist, however, and morphological thought cannot with impunity ignore them.

The urge of the phyllome towards whole-shoot characters, for which we are offering evidence, is revealed most convincingly in the more highly elaborated leaves, while those of simplest form (e.g. scale-leaves) show little hint of shoot structure. In general the aspiration towards whole-shoot-hood is more completely obvious in certain compound-pinnate leaves than in the peltate leaves which we have been considering. The rachis in the compound-pinnate leaf may be held to correspond to the axis of the shoot, and the pinnae to the leaves, while the attachment levels of the pinnae recall shoot-nodes, and may be distinguished as 'nodelets'. The arrangement and stance of the pinnae in pinnate leaves is significant from our standpoint. Just as the two-ranked leaves of a shoot may either be opposite (e.g. *Potamogeton densus* L.) or alternate (e.g. Gramineae), so the two rows of pinnae of a compound leaf may either be opposite (e.g. *Angelica ampla* Nelson, Fig. 18, B1, p. 115, and *Melianthus major* L., Fig. 18, D1), or alternate (e.g. *Lapsana communis* L., Fig. 18, C1). In ordinary bifacial pinnate leaves, the placing of the pinnae is lateral (i.e. lengthwise along the rachis), since they arise as outgrowths from the two edges of the rachis, that is to say, from the intersection of the upper and lower surfaces of the phyllome.[1] The pinnae thus all lie in the plane of flattening of the leaf; but, in cases where the rachis tends towards the axial

[1] See pp. 79, 80.

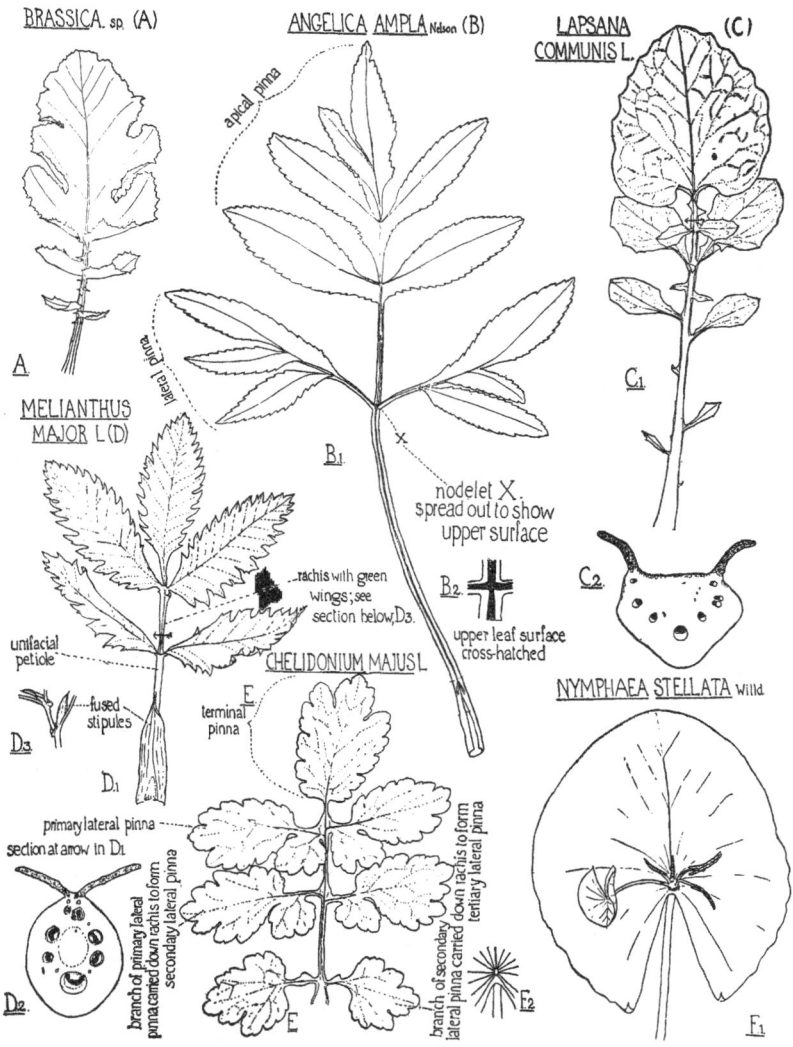

Fig. 18. All drawings (× $\frac{1}{6}$ *circa*), except B2 (on larger scale), and C2, D2 (×6 *circa*).

character of unifaciality, the position of the pinnae changes. When the rachis is completely unifacial, i.e. when its margins are in contact along the adaxial median line, these margins separate only for a brief space at each nodelet, and then close again, with the result that the exsertion of the pinnae becomes transversal.[1] This is illustrated in the lower pair of pinnae in *Angelica ampla* Nelson (Fig. 18, B 1, p. 115). These pinnae are held in a plane almost at right angles to the phyllome surface, while the upper lobes are purely marginal and lie in the general plane of the leaf. The closer the approach to radiality in the rachis, the more the relation to it of the pinnae comes to resemble that of leaves to an axis; and the leaf as a whole thus approximates to a shoot. A highly organised form of this shoot-like scheme of rachis and pinnae is found in the umbellifer, *Carum verticillatum* Koch,[2] in which the segments of the leaflets stand out from the rachis like a tufted whorl, the appearance of which must have suggested the specific name. The leaf as a whole resembles a shoot of *Ceratophyllum*, in which the whorling is again apparent only; but whereas it is due in *Ceratophyllum* to the sub-division of a pair of opposite *leaves*,[3] in *Carum* it comes about through sub-division of a pair of . opposite *pinnae*. *Carum* shows another shoot-like character in the long-continued growth of the phyllome tip, which may result in a leaf over 40 cm. in length;[4] as many as 36 'internodelets' have been recorded.[5] Prolonged meristematic activity in the apical region is found in various compound-pinnate leaves; that of the Caucasian wing-nut (*Pterocarya fraxinifolia* K. Koch), for instance, may have 22 leaflets, and the tip, before it reaches maturity, consists of a group of young pinnae, recalling the terminal bud of a shoot (Fig. 19, A 1, A 2, p. 117). Such leaves remind one of those of certain ferns which, after many years of life, still have a rolled tip, retaining the power of apical growth.[6] There are other large pinnate leaves among dicotyledons which, in their size and in the number of their pinnae, suggest luxuriant shoots; the leaf of *Ailanthus Giraldii* Dode, var. *Duclouxii* Dode, may attain, under garden conditions in this country, a length of 81 cm., and bear as many as 38 leaflets.

[1] Troll, W. (1934*a*), fig 10, p. 92.   [2] Troll, W. (1934*a*), p. 93.
[3] Schaeppi, H. (1935).   [4] Troll, W. (1934*a*), p. 93.
[5] Glück, H. (1911), table, p. 330.   [6] Christ (Christ-Socin), H. (1897), p. 74.

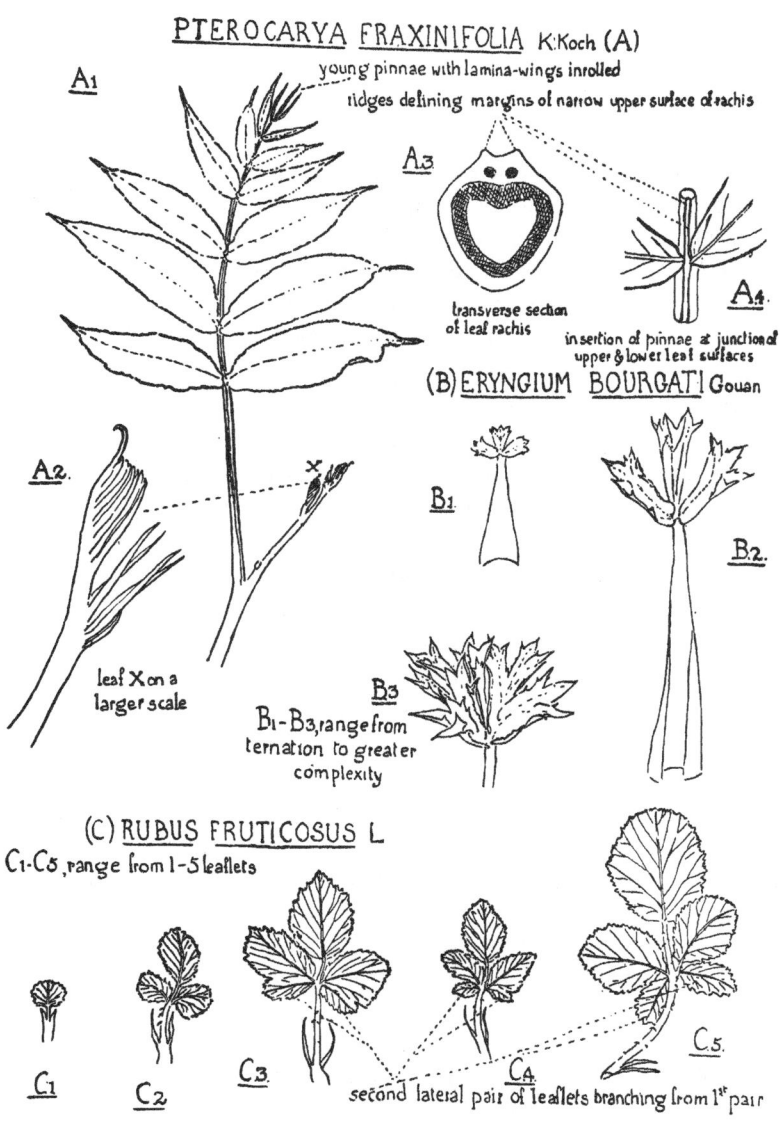

PTEROCARYA FRAXINIFOLIA K:Koch (A)

A1

young pinnae with lamina-wings inrolled

ridges defining margins of narrow upper surface of rachis

A3

transverse section
of leaf rachis

A4

insertion of pinnae at junction of
upper & lower leaf surfaces

(B) ERYNGIUM BOURGATI Gouan

A2.

x

B1

B2.

leaf X on a
larger scale

B3

B1-B3, range from
ternation to greater
complexity

(C) RUBUS FRUTICOSUS L

C1-C5, range from 1-5 leaflets

C1

C2

C3.

C4.

C5.

second lateral pair of leaflets branching from 1st pair

Fig. 19. All drawings (×½) except A 2 (larger scale than A 1)
and A 3 (× 14 *circa*).

117

A minor point of similarity between compound-pinnate leaves and shoots is that the rachis of the pinnate leaf may be winged by the bases of the pinnae (e.g. *Melianthus major* L., Fig. 18, D 1, D 2, p. 115), as a shoot-axis may be winged by the bases of the leaves. The similarity is particularly striking in such a shoot as that of *Acacia alata* R.Br. (Fig. 7, A, p. 88), to the dorsiventrality of which we have already referred.[1]

To the comparison of compound-pinnate leaves and shoots, it may be objected that there is an essential difference in the order of development of the appendages, the leaves of a shoot being, as a rule, borne acropetally, while the pinnae of a pinnate leaf may be produced in a basipetal series. This, however, is perhaps merely one of the divergences due to the contrast between the continued apical growth of the shoot, and the abbreviated character of leaf growth, in which the apex remains meristematic for a relatively short time. As a compensation for this apical limiting, there is a tendency for a pinna or leaf-lobe to develop below the apex on either side, thus producing the ternation which is so frequent a feature of the phyllomes of angiosperms. Various leaves, which at maturity reach a more complex form, pass through a definite ternate stage in early development; this is true of the horse-chestnut (*Aesculus Hippocastanum* L.), in which the remaining pairs of leaflets arise later through branching of the primary lateral pair.[2] Moreover, when we turn to non-ontogenetic evidence, we find many examples of simple leaves, in which two conspicuous main lateral veins occur in addition to the midrib, and of compound leaves with three leaflets, one median and two lateral. Ternation, as a first step towards greater elaboration, is illustrated by the successive seedling leaves of *Rubus laciniatus* Willd. (Fig. 21, B 1, p. 121), and by other examples in our figures, such as *R. fruticosus* L. (Fig. 19, C 1–C 5, p. 117), and *Eryngium Bourgati* Gouan (Fig. 19, B 1–B 3, p. 117). In the ash (*Fraxinus excelsior* L.) the seedling (Fig. 20, A, p. 119) shows a series beginning with simple leaves, which are followed by leaves with two additional pinnae below the terminal one, while this ternate phase is succeeded by further degrees of pinnation, and even, though rarely, by a compound-pinnate form (Fig. 20, B and C). Con-

[1] See p. 87.   [2] Foster, A. S. (1929).

118

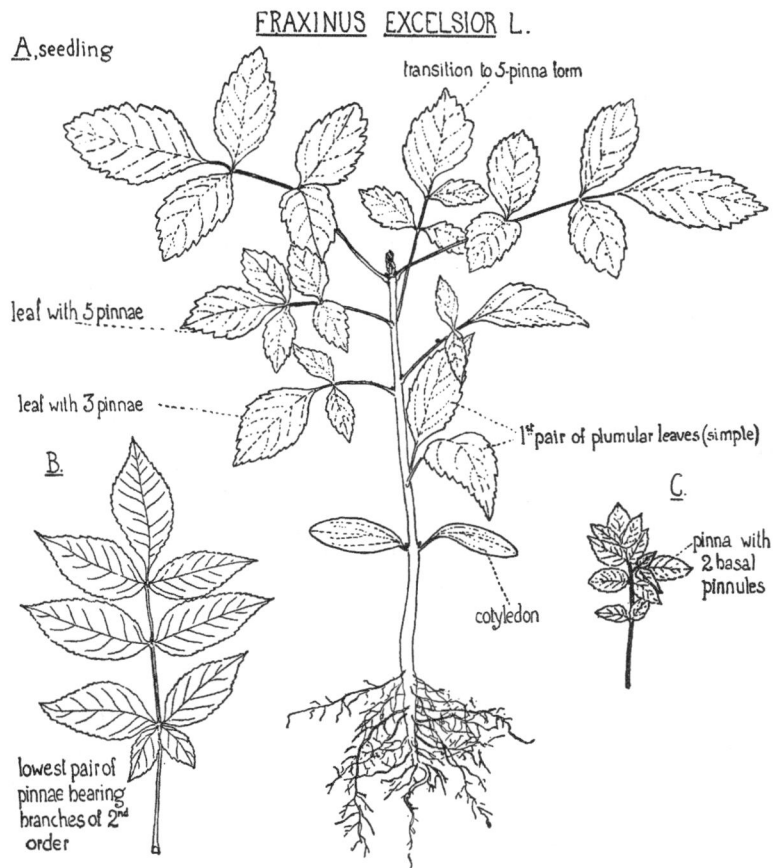

FRAXINUS EXCELSIOR L.

A, seedling

transition to 5-pinna form

leaf with 5 pinnae

leaf with 3 pinnae

1st pair of plumular leaves (simple)

B.

C.

pinna with 2 basal pinnules

cotyledon

lowest pair of pinnae bearing branches of 2nd order

Fig. 20. *Fraxinus excelsior* L.: A, seedling showing succession of leaf forms from simple, through ternate, to 5-pinnate ($\times \frac{1}{3}$); B, leaf in which each member of the basal pair of a 7-pinnate leaf bears a secondary pinnule thus producing a 9-leaflet form ($\times \frac{1}{3}$); C, small leaf in which one unpaired pinna bears two basal pinnules, thus producing a 12-leaflet form ($\times \frac{1}{3}$).

versely, in the flowering shoot of the blackberry (*Rubus fruticosus* L.), the mature leaf with its five pinnae is reduced first to a ternate and then to a simple form. In ternation, followed by branching, we have, indeed, the key to the interpretation of the pinnate leaf.[1] Such a phyllome as that of *Angelica ampla* Nelson

[1] On the Umbelliferae see Uittien, H. (1928*a*) and (1928*b*).

(Fig. 18, B, p. 115), for instance, clearly represents an elaboration of the three-leaflet scheme.

When we try to understand how the various forms of pinnate leaf may be visualised, one of the general characteristics of pinnae becomes significant. This is the asymmetry, which they not infrequently show, corresponding with the asymmetry often seen within the lamina-half of a simple leaf.[1] Usually the side towards the leaf-base is the one that is the more developed,[2] as in the *Epimedium* sketched in Fig. 25, p. 138; such asymmetry, whether in a pinna or a whole leaf, may be regarded as a first hint of possible branching. After the primary ternation, such branching may be carried out, either by the apical pinna itself, or by the paired laterals. These alternatives may be illustrated on comparing a suite of leaf specimens of the loganberry (? *Rubus Idaeus* L. × *R. fruticosus* L., Fig. 21, C 1–C 4, p. 121), and of the blackberry (*R. fruticosus* L., Fig. 19, C 1–C 5, p. 117). In the loganberry, the change from three to five leaflets comes about through a second basal branching of the apical pinna, while in the blackberry, on the other hand, this change is the result of basal branching of the first lateral pair. A comparable difference exists between certain leaves of *Medicago lupulina* L., which develop an extra pinna; Fig. 21, D 2, p. 120, represents an attempt, as it were, at the loganberry type, and Fig. 21, D 3, at the blackberry type. If the apical pinna of any leaf were to continue the process as after another, the result would be a pinnate leaf in which the pinnae (excluding the terminal pinna) would be arranged acropetally. If, on the other hand, the process indicated in the blackberry were continued—that is to say, if the branching were delegated by the apical pinna to the paired laterals, so that they each branched once, on the basal side, and then handed on their function to their daughter pinnae, and so on, through successive generations, the result would be wholly basipetal. If this basipetal development takes place without separation of the leaflets, we arrive at such pedate leaves as those sketched in Fig. 21, A

[1] Cf. *Tilia*, pp. 176 (Fig. 38), 177.
[2] In certain rare cases the distal halves of pinnae exceed the basal; this can be recognised in the pinnae and pinnules of *Gleditschia caspica* Desf. (Frontispiece, B 1 and B 3), though here it is not very conspicuous. For a full account of this type of asymmetry in the leaves of Meliaceae, see Briquet, J. (1935–6).

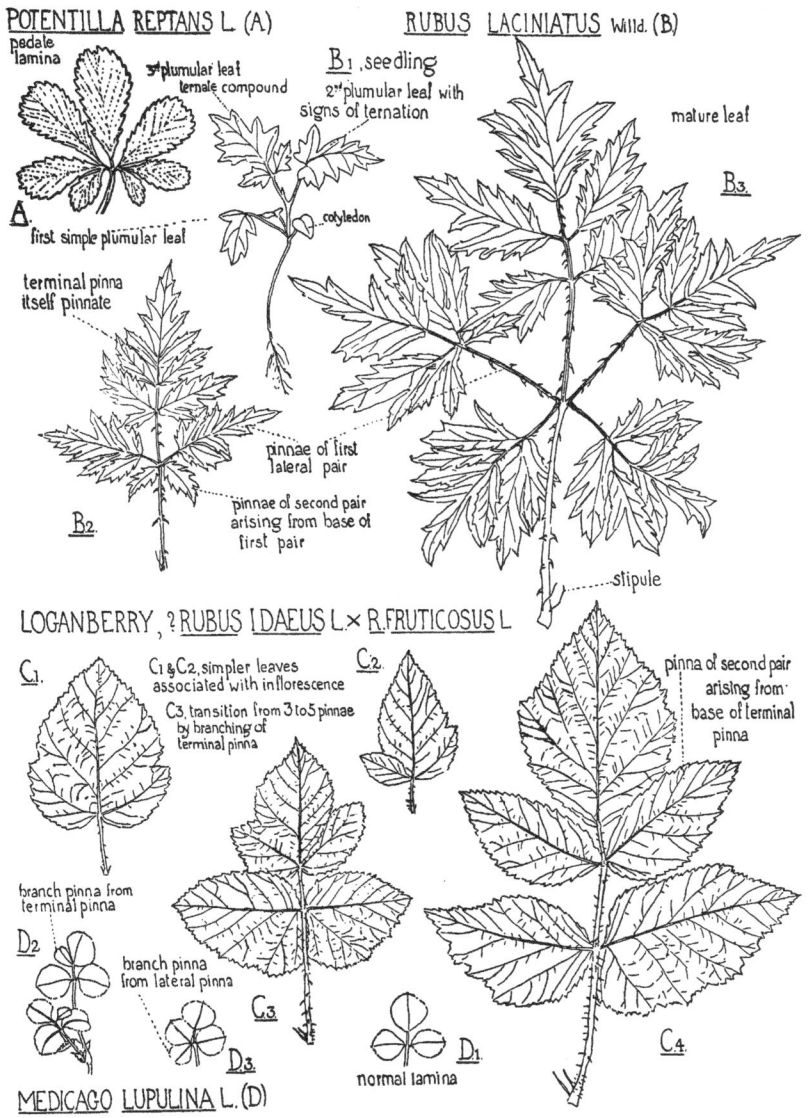

Fig. 21. All drawings (× ⅓) except B 1 (× ⅔).

121

and B, p. 121; the compound-pinnate leaf is reached, on the other hand, by the 'stringing out' of the leaflets along the rachis by means of its intercalary growth. In Fig. 18, E, p. 115, such an

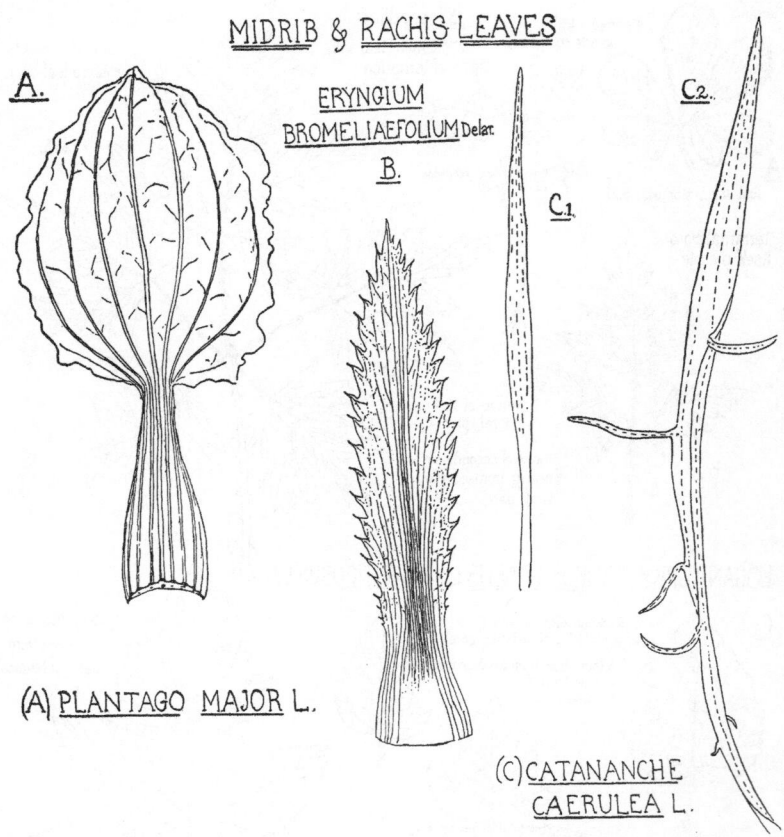

MIDRIB & RACHIS LEAVES

A.

ERYNGIUM
BROMELIAEFOLIUM Delar.

B.

C1.

C2.

(A) PLANTAGO MAJOR L.

(C) CATANANCHE
CAERULEA L.

Fig. 22. A, *Plantago major* L., lamina formation by lateral separation of veins ($\times \frac{1}{2}$). B, *Eryngium bromeliaefolium* Delar., small leaf ($\times \frac{1}{2}$). C, *Catananche caerulea* L. var. *alba*: C1, leaf which had an inflorescence in its axil ($\times \frac{1}{2}$). B and C2 may be regarded as intermediates between rachis leaves and pinnate leaves.

interpretation is indicated for *Chelidonium majus* L. Examples in which the existence of 'stringing out' can be demonstrated more readily than in ordinary pinnate leaves, may be found among phyllomes with relatively small intermediate leaflets (*Zwischen-*

fiedern),[1] such as those of *Agrimonia Eupatoria* L. (Fig. 13, D, p. 103); these miniature pinnae caught the attention of the artist of the *Codex Aniciae Julianae* of Dioscorides more than fourteen centuries ago.[2]

We may think of the contrast between shoot development (acropetal growth), and compound-pinnate leaf development (ternation with serial derivative growth), as a correlative of the apical continuity of growth in the shoot, and the limitation of growth in the leaf. In the shoot, the apex has a continuous meristematic life, and thus produces successive outgrowths (leaves) in acropetal order; here the 'stringing out' separates these leaves as wholes. In the leaf, on the contrary, the tip can do little, as a rule, in forming appendages before its growth period ceases, so the capacity for further development suffers a compensatory transference to the pinnae (where it may give rise to a basipetal scheme), and also to the non-apical part of the rachis (where it takes the form of intercalary elongation); as a result, the 'stringing out' detaches basal branches from the outgrowths (pinnae). In certain exceptional cases, on the other hand, by compensatory growth in width instead of length, the midrib or rachis may expand laterally to form a lamina[3] (Fig. 22, p. 122).

[1] Müllerott, M. (1940). A sequence of leaf outlines of different varieties of potato, in Salaman, R. N. (1946), figs. 4–25, pp. 17–20, though drawn without reference to any theory of the nature of *Zwischenfiedern*, yet affords evidence for the correctness of Troll's view, as developed by Müllerott, according to which these structures are pinna-branches separated from the parent pinna by intercalary growth of the leaf rachis.

[2] Gunther, R. T. (1934), black and white sketch of this drawing on p. 222. On the *Codex Aniciae Julianae*, see Arber, A. (1938*a*), pp. 8–10, 185, etc.

[3] See p. 85.

CHAPTER VIII

THE BEARING OF THE PARTIAL-SHOOT THEORY OF THE LEAF ON OTHER MORPHOLOGICAL PROBLEMS

IN previous chapters we have discussed the more direct evidence for the partial-shoot theory of the leaf; it remains to assess its claims on somewhat different lines, by considering whether it will throw light on certain outstanding morphological problems.

One of these problems, though a minor one, is that of the terminal phyllome. Those who regard the leaf as belonging to a category, or type, completely distinct from that of the stem, generally maintain that a leaf, being defined as a lateral member, *cannot* terminate a shoot; but they are faced with the difficulty that, in certain cases, the apex of an inflorescence axis is, in fact, wholly transformed into a terminal bract.[1] A carpel, moreover, may be clearly terminal to a floral axis; this has been demonstrated, for instance, on ontogenetic evidence, for the almond (*Prunus Amygdalus* Stokes),[2] while the floral anatomy shows it to be true also for *Mirabilis Jalapa* L. On the partial-shoot theory of the appendages, such terminalness presents, however, no difficulty. There is no reason why, as a final effort before cessation of all further development, a shoot apex should not be wholly used in the construction of something which is of the same nature as itself, though lacking completeness—in other words, in the formation of a partial-shoot (phyllome). The same thing happens within the phyllome itself, for the terminal pinna in an imparipinnate compound leaf bears the same relation to the rachis that a terminal leaf bears to the axis. It seems then that, in the light of the partial-shoot theory, the 'problem' of the terminal leaf proves to have no real existence.

---

[1] Arber, A. (1937a); references on p. 162.   [2] Brooks, R. M. (1940).

A question of greater significance, which suggests itself in relation to the partial-shoot theory, is that of the nature of the axillary bud. It is generally agreed that the vast majority of buds arise in connexion with an 'axillant' leaf; but to what, exactly, this connexion amounts, remains vague, and Warming's criticism, though made in the eighteen-seventies, is still pertinent. "Personne," he wrote, "ce me semble, n'a encore exprimé en termes clairs et précis quelle est la relation entre un bourgeon et la feuille dite feuille-mère. En général, on rencontre seulement des expressions peu précises comme celles-ci, que les bourgeons sont situés 'dans l'angle' entre la feuille et l'axe-mère, ou 'à l'aisselle' de la feuille."[1] The conclusion that Warming drew from his studies of the development of bud and leaf—studies which were most remarkable, considering the state of botanical technique at that date—was that the axillary bud belonged quite as much to the axillant leaf as to the parent axis, and that it was, indeed, in great part a development of the leaf-base. Modern work has confirmed this notion in certain instances; Majumdar[2] for instance, has described the axillary buds of the umbellifer, *Heracleum Sphondylium* L., as produced entirely within the tissue of the leaf or leaf-cushion. Warming held also that the axillant leaf might be treated as the first and only leaf of the axillary bud, while the succeeding leaves, apparently belonging to this bud, are really borne by the buds of the next order. Suggestions of this unorthodox type do not seem at first glance to offer a consistent scheme, but they may be integrated if we think of them as incomplete and tentative expressions of a generalisation which was also formulated by Warming—namely, that *the axillant leaf and its axillary bud together constitute a unit.* It remains to be seen whether any affinity can be traced between this view and the partial-shoot conception of the leaf; for, if these two theories each represent a certain aspect of the truth, it should be possible to harmonise them.

We have seen that, on the partial-shoot theory, a typical leaf is a shoot in which the apex is limited in its power of elongation and in its radiality. The failure in lengthening arouses a tripartite character in the leaf, because the frustrated growth of the tip seeks to compensate itself by diversion into a basal branch on

[1] Warming, E. (1872).   [2] Majumdar, G. P. (1942).

either side;[1] while the failure in radialness, brought about by inactivity on the adaxial face of the tip, induces a dorsiventral character, though a residual urge towards the radiality of whole-shoot-hood also persists. As a corollary to the partial-shoot theory, it may be suggested that the axillary bud should be regarded as a direct outcome both of the tripartedness and of the urge to radiality, which are such fundamental features of the leaf. Envisaged on these lines, *the axillary bud may be pictured as equivalent to two branches (lobes or pinnae) produced from the bas-al region of the axillant leaf, on either side; these branches must be visualised as in a state of face-to-face fusion in front of the parent leaf. Though they each, as segments of the leaf, would be of dorsiventral organisation, they achieve radiality by their union.*

On the partial-shoot theory, with its axillary bud corollary, the above-ground part of the plant thus consists of closely linked series of whole- and partial-shoots. The plumular whole-shoot bears leaves (partial-shoots), each of which may give rise to an axillary bud (whole-shoot), which again produces partial-shoots; and this alternation of whole- and partial-shoots may proceed indefinitely. The partial-shoot, though unable itself to reach whole-shoot level, yet achieves whole-shoot-hood in its offspring, the lateral shoot. To obtain actual 'proof' for such a theoretic conception as this of the axillary bud, can scarcely be hoped; the question rather is whether it is consistent with the characteristic facts of the structural relation between leaf and axillant shoot. We find that, though there is much variation in the mode of vascular connexion between these members, it is yet possible to offer a statement about it which has some degree of generality. According to de Bary's[2] summary, in the majority of those dicotyledons and gymnosperms in which the stem in transverse section shows a ring of vascular bundles, the axillary bud is, as we have already mentioned in another connexion,[3] supplied by two strands. These ramular bundles, or 'bundles of insertion', arise from the margins of the gap left in the cylinder by the median strands of the leaf. The pair of bundles face one another, xylem towards xylem, and, in supplying the axis of the daughter shoot, they each divide into an arc, so that their products jointly form a ring. Examples can readily be found, both among angio-

[1] Cf. pp. 97, 98.   [2] Bary, A. de (1884), p. 307.   [3] See pp. 88 (Fig 7, B), 91.

sperms[1] and gymnosperms.[2] It seems reasonable to suppose that this recurrent scheme, despite the many exceptions that could be cited, has some significance as an indication of the morphology of the parts which the bundles serve. Now this bundle relation is precisely what might have been foreseen on the interpretation of the axillary bud here proposed; for the face-to-face approach of the two bundles arising from the sides of the leaf-gap, and their ensuing divisions to produce a radial system from two dorsiventral systems, are completely compatible with the fusion of a basal pair of leaf branches, which we have pictured as responsible for bud formation.

Our theory of the bud, at least in its present form,[3] is intended to apply only to those that are axillary. It may be recalled, however, that certain buds, which seem at first glance not to belong to this category, may, in fact, fall into the class of axillary structures. In this connexion it is helpful to consider the 'bractless' raceme of the crucifers, in which the axillant leaf, though generally supposed to be non-existent, may, in fact, be present, in the attenuated form of two ephemeral and non-vascular stipules.[4] If basipetal reduction of the leaf-members continued, the next stage would be the complete suppression even of the stipules, so that nothing would be left of the *leaf-bud* unit, except the bud, which then would be apparently non-axillary.

A knotty problem, closely bound up with the interpretation of the axillary bud, is that of the nature of the female reproductive shoot in the conifers. The cone axis bears a number of complex units, each consisting of an external sterile phyllome (bract), with a fertile member (ovuliferous scale, *Fruchtschuppe*) situated between the bract and the cone axis. This arrangement is very ancient; it has been demonstrated for a *Voltzia* from the

[1] For a description and a very clear three-dimensional diagram, see Miller, H. A. and Wetmore, R. H. (1946), p. 3, and fig. 4, D, p. 4. Cf. also Arber, A. (1931*b*), strands for flower 3, fig. 2, D 1, p. 321, and fig. 3, A 1–A 5, p. 323; and (1932), fig. 1, A 3–A 5, p. 147; Dormer, K. J. (1945), p. 141; Philipson, W. R. (1947*a*), p. 293.

[2] De Bary's account is in accord with earlier and later work; see Geyler, H. T. (1867); Robertson (Arber), A. (1906), p. 260, confirming Geyler, H. T. (1881); Brooks, F. T. and Stiles, W. (1910), p. 312; Eames, A. J. (1913), p. 21; Aase, H. C. (1915), pp. 285–7; Sahni, B. (1920), p. 261.

[3] The writer fully realises the tentative and provisional character of this theory, and hopes to develop and test it later by a more extended study.

[4] Cf. p. 99.

Permian.[1] In modern conifers, the bract and ovuliferous scale may be distinct, or fused in varying degrees. There has been little difference of opinion about the 'bract', which is manifestly foliar: the crux is the interpretation of the ovule-bearing scale. This has been the subject of an active controversy, waged throughout the second half of the nineteenth century, and still continuing to-day. Its content has been summarised repeatedly,[2] so it is unnecessary here to do more than to analyse, on broad lines, the main solutions that have been offered, and to relate them to our particular standpoint. The various theories proposed fall, roughly, into two categories, according to whether the ovuliferous scale is regarded as (1) a short, leaf-bearing shoot, axillary to the bract, so that the cone, as a whole, is an inflorescence; or (2) a part of the bract itself, so that the bract and ovuliferous scale together compose one leaf, and the whole cone is a single flower. The majority of botanists have adhered to the short-shoot theory, formulated by Braun[3] nearly a hundred years ago. Among those accepting this theory, there is a general consensus of opinion as to its main outlines; but exactly how the individual parts of the ovuliferous scale are to be equated in detail with those of a leafy shoot, is open to different interpretations; the result is that, under the short-shoot theory, there are ranged a whole series of divergent subsidiary hypotheses. These, however, do not affect our present discussion; we need concern ourselves here only with the widely held belief that the ovuliferous scale is of shoot nature.

Convincing evidence for the short-shoot hypothesis can be drawn from Sahni's account of *Acmopyle*,[4] and from Hagerup's[5] more recent description of the unit of cone structure in various forms; both these writers base their conclusions on an exhaustive study of serial sections. In *Cryptomeria japonica* (L.-f.) Don, for instance, Hagerup finds that the ovuliferous scale arises as a wart-like secondary axis, given off from the main axis of the cone. This protrusion reveals its axial character in the radial construction of its stele. Moreover, as Sahni and other

[1] Walton, J. (1929).
[2] See especially Pilger, R. (1926); also, Worsdell, W. C. (1900); Coulter, J. M. and Chamberlain, C. J. (1917); Chamberlain, C. J. (1935).
[3] Braun, A. (1853), p. 81, footnote.   [4] Sahni, B. (1920).
[5] Hagerup, O. (1933).

writers[1] have shown, the bundle system of the ovuliferous scale follows the same general scheme as that of the vegetative branch, for the ovuliferous scale is supplied by two converging strands, arising one on either side of the bract bundle.

Those who are opposed to the short-shoot theory, hold views based on the idea that the bract and ovuliferous scale together constitute one leaf. The scale has sometimes been regarded as an enation from the 'bract,'[2] while it has also been suggested that the bract and scale result from a serial splitting of one member, a process which can be paralleled in the cones of the Sphenophyllales.[3] These hypotheses may be considered as variants of that of Sachs,[4] who wrote, long ago, that he was inclined to regard the ovuliferous scale, not as axillary, but as merely a strongly developed placenta arising from the bract.

Sachs's theory was given a slightly different turn, and a wider application, by Delpino,[5] whose work was based upon a study of the vegetative twinned and united needles of *Sciadopitys* (the umbrella pine) which appear to be borne in the axils of scale leaves. According to Delpino, however, *the scale-leaf and double needle together* are equivalent to one phyllome, the needle representing paired basal segments, which arise from the scale-leaf and are fused in front of it. This theory seems to have been ignored by many of the more recent writers, who have generally held the double needle to be a pair of leaves borne on a minimal short-shoot, axillary to the scale leaf. Delpino's idea about *Sciadopitys* led him to apply corresponding reasoning to the ovuliferous scale of the Abietineae and other conifers.[6] He regards the so-called bract as the median region of a potentially tripartite carpel. This region is sterile, while there is a fertile placental segment on either hand; these two segments fuse in front of the sterile median region to form the ovuliferous scale. Delpino shows that his view brings the Abietineae into line with the angiosperms, in which he recognises each carpel as an essentially tripartite

[1] Strasburger, E. (1872); Brooks, F. T., and Stiles, W. (1910), p. 312 (peduncle supply); Eames, A. J. (1913), p. 21; Sahni, B. (1920).
[2] E.g. Thomson, R. B. (1940).   [3] Hirmer, M. (1932).
[4] Sachs, J. von (1868), p. 427.
[5] Delpino, F. (1889*b*); his views are also summarised very shortly in Penzig, O. (1921, 1922), vol. III, pp. 507–8.
[6] Delpino, F. (1889*a*, *b*).

member, the two laterals (placentae) either remaining free from one another; or fusing into a single body opposite the sterile median region; or becoming united, in various ways, with adjacent carpels.

Having sketched in barest outline the existing opinions as to the nature of the ovuliferous scale in the conifers, it remains to consider these views in relation to one another, and to the theory of the axillary bud, with which we are here concerned. The most striking fact that emerges from a review of the literature, is that, despite a hundred years of discussion, neither the shoot interpretation nor the leaf interpretation can be said to have displaced its rival. The arguments for both are cogent, and it is significant that a botanist, even of such insight and experience as Eichler, should be found to have ranged himself first on one side and then on the other.[1] The obvious deduction from this state of things is that both the hypotheses represent aspects of the truth, and that, instead of treating them as mutually exclusive, a standpoint should be sought from which they can be synthesized; and it may be suggested that the present theory of the axillary bud does, in fact, serve this purpose. Returning to the comparatively simple case of the *Sciadopitys* 'double needle', we see that, if we are prepared to adopt the view that, in general, the axillary bud represents a pair of fused basal outgrowths from its 'axillant' leaf, Delpino's theory (that the needle of *Sciadopitys* consists of a pair of fused outgrowths from the base of the scale) becomes merely an alternative method of expressing the more orthodox opinion (that it is formed from a minimal short-shoot, axillant to the bract, and bearing a pair of united leaves). If this is true of the *Sciadopitys* needle, it may equally well be true of the cone scale of the Abietineae. The antithesis between the short-shoot and the foliar theories of the ovuliferous scale would then be resolved, for, if all axillary buds represent fused leaf-lobes, and thus all lateral shoots are in their origin foliar, the shoot and the leaf theories are different forms of expression for the same truth, and their apparent antagonism becomes no more than a matter of terminology. It may be objected that this synthetic attitude ignores the distinction that, on the shoot theory, the cone is considered to be an inflorescence, whereas, on the leaf theory, it

[1] Cf. Eichler, A. W. (1863) and (1881), p. 1031.

is a single flower; but, if the present concept of the axillary bud be adopted, there is no longer the same rigid demarcation between flower and inflorescence. In the next chapter, looking at the matter from another point of view, we shall find that, correspondingly, the flower *Gestalt* in different plants may find expression in different orders of branching.[1]

The axillary bud theory may also be applied, though less directly, to the carpel of the angiosperm. This member is generally regarded as purely foliar; according to Delpino's view, it is, as we have seen, analysable into a median sterile lobe and two lateral fertile lobes (placentae). Some botanists, however, have considered certain placentae to be axial.[2] This difference of opinion would lose its significance if we suppose that the fertile lobes, by mutual fusion, may together rise to the radiality of a whole-shoot, and thus form a structure comparable to an axillary bud.

It is perhaps possible that another antithesis—that between the shoot and the leaf interpretations of the ovule[3]—will find its resolution on lines parallel to those suggested for the problems of the double needle of *Sciadopitys*, the cone scale in the Abietineae, and the angiosperm carpel.

A review of the considerations brought together in this chapter shows that the axillary bud theory offers a means of fusing into inclusive forms certain pairs of opposed views, which have hitherto been regarded as mutually exclusive. Its capacity for synthetic application indicates that both it, and the partial-shoot theory of the leaf, of which it is a corollary, possess some degree of validity.

In thinking about the plant body, we have so far confined attention to the shoot—that is to say, essentially to the above-ground region—and we have left the problem of the nature of the root system untouched. We must now enquire if this problem is in any way illumined by the partial-shoot theory of the leaf.[4]

In the early days of herbalism, the underground part of the plant attracted a good deal of interest, because of its real or supposed medicinal qualities; but, since concern with this aspect

[1] See pp. 143 *et seq.*

[2] For a recent presentation of this view, see Hagerup, O. (1942).

[3] See also p. 142.

[4] The theory of the root suggested in the present chapter, was proposed in Arber, A. (1941 *a*).

has waned, the attitude of botanists to the root has become step-motherly. This is perfectly natural, since, owing to its burial in the earth, the study of the root is fraught with inconvenience; but the resulting neglect is regrettable, as it fosters a tendency to think of the plant, in relation to the environment, in terms of the above-ground parts alone. Actually, however, this relationship is more extensive and intimate in the case of the root than the shoot. In a plant of rye (*Secale cereale* L.), for instance, which was fully investigated from this point of view, the surface area of the subterranean parts proved to be 130 times that of the above-ground shoots.[1] It may well be that a certain monotony, that is conspicuous in root- as compared with shoot-structure, is to be attributed to the interment of the root in a medium which suppresses its freedom to develop its own potentialities.

In various respects shoot and root show a clear parallelism. They both share radial symmetry and apical growth. Moreover they are both capable of giving rise to fresh units of their own kind (respectively, lateral shoots and lateral roots), and also of their partner's kind (respectively, adventitious roots arising from shoots, and shoot-buds arising from roots,[3] such as those shown for the elm in Fig. 23, p. 133). The tendency, also, to domination of the main axis by branches of later generations, which we have traced in the shoot, is repeated in root systems; in many plants the main root dies away at an early stage, and is replaced by adventitious roots. Though such resemblances between shoot and root are significant, the differences between these two organs are sufficiently salient to make it difficult to find any unified concept which shall include them both. Goethe observed that he had the same respect for the root of the plant as he had for the foundations of Cologne Cathedral, but that he felt that an organ, showing endless variety but no progression, was nothing to him.[2] Elsewhere, however, he threw out the suggestion that the root is a leaf which absorbs moisture beneath the earth; he thus brought it under his hypothesis that "Alles ist Blatt".[4] If now we use the partial-shoot theory, as Goethe used his leaf concept,

[1] Dittmer, H. J. (1937).
[2] Goethe, J. W. von (1887 etc.), Abt. II, Bd. 6, *Zur Morphologie. Verfolg. Unbillige Forderung.* (1824), pp. 331–2.
[3] Cf. Rauh, W. (1937).
[4] Troll, W. (1926), p. 52.

can we arrive at any corresponding view of the root? According to our thesis, the leaf is a partial-shoot, revealing an inherent urge towards becoming a whole-shoot, but never actually attaining this goal, since radial symmetry, and the capacity for apical

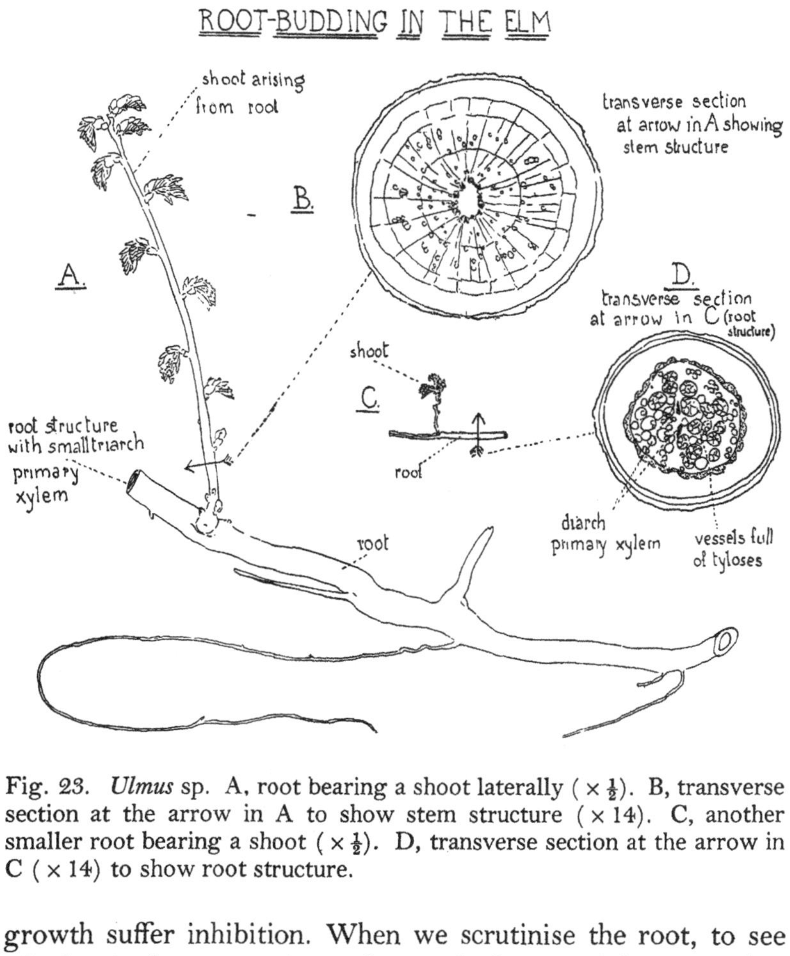

ROOT-BUDDING IN THE ELM

Fig. 23. *Ulmus* sp. A, root bearing a shoot laterally (× ½). B, transverse section at the arrow in A to show stem structure (× 14). C, another smaller root bearing a shoot (× ½). D, transverse section at the arrow in C (× 14) to show root structure.

growth suffer inhibition. When we scrutinise the root, to see whether it shows any signs of an equivalent partialness, we find that it may occasionally be dorsiventral, thus imitating, as it were, the partial-shoot character of the leaf. The 'thallus' of certain members of the Podostemaceae[1] affords a striking example, but such cases are rare. In general, the differences

[1] Willis, J. C. (1902), see especially pp. 406–23.

between the root, and the partial-shoot which takes leaf form, lie in the fact that the root retains the two whole-shoot-like features—radial symmetry, and apical growth—of which the leaf is to a great extent deprived. These, however, are the very deprivations to which the leaf owes its partialness in comparison with the shoot. Is it possible to regard the root, also, as a partial-shoot, but one that derives its partialness from *a different set of deprivations* from those suffered by the leaf? A clue to what is essential in root nature may perhaps lie in the fact that both lateral and adventitious roots tend to arise deep in the plant body, the surface tissues playing no part in their origin. We know from studies of periclinal chimaeras, that there may be a certain reality in the distinction between external and internal regions, since root cuttings may give rise to plants whose characters are those of the internal component alone of the chimaera.[1] It has been suggested by Thoday[2] that the rhizophore of *Selaginella* might, from a certain standpoint, be regarded as "a sort of chimaera, with a core of root inside a skin of shoot". It is possible that this suggestion might be extended to cover a much wider field. The hypocotyl of the flowering plants obviously tends to combine external shoot characters with internal root characters; and there is something to be said for visualising the shoot as a whole on corresponding lines. May we then compare the shoot in general to a periclinal chimaera, of which the inner component is of root nature? A similar suggestion, though couched in medieval terms, was made long ago by Albertus Magnus,[3] who regarded the root as continued up into the interior of the shoot, where it formed the pith or core. The same idea reappeared some 500 years later in the writings of Linnaeus.[4] This interpretation of the root is consistent with its inability to produce sporogenous tissue, which arises, normally, from the more superficial layers of the plant body.

The urge towards whole-shoot characters, which we have recognised in the leaf, may be detected, though less frequently, in the root. The root-thallus of the podostemads sometimes shows remarkably shoot-like features; it has, indeed, been pointed out that the foliaceous root-thalli of certain species of *Dicraea* are not unlike the prostrate secondary shoot-thalli in a species of

---

[1] Bateson, W. (1916), (1921). [2] Thoday, D. (1939), footnote, p. 101.
[3] See pp. 25, 26. [4] Linnaeus, C. (1751), § 157, p. 99.

Oenone.[1] Apart from such peculiar cases as these, the urge, which we are postulating, reveals itself in the anatomy of roots in general. Different as the primary scheme of tissues of root and shoot may be, the secondary thickening of the root often produces an internal structure deceptively like that of an old stem.

All this evidence is admittedly slender, but it at least suggests the possibility of interpreting root as well as leaf as a partial-shoot. The entire plant body would thus consist of shoots and partial-shoots, so that we reach a unified conception of the construction of the plant as a whole. Another and more general advantage of this theory is that, when we think of root and leaf as being both, in their different ways, partial-shoots, we are envisaging the simpler as logical (though not temporal) derivatives of the more complex. To see the matter in this directional sense, rather than vice versa, accords with the trend of modern thought, which inclines more and more to the interpretation, for instance, of the inorganic in terms of organism, rather than to the mechanistic explanation of the complex living thing in terms of its simpler non-living components, according to the mental mode of the nineteenth century. Another respect, in which the partial-shoot concept is at home in the intellectual atmosphere of to-day, is in the stress which it lays upon the dynamic study of relations and of parallels, rather than upon the static study of fixed types and absolute categories. That the partial-shoot theory, with its corollaries, is open to criticism in many directions is, however, obvious;[2] the utmost that can be hoped is that it will be a temporary expedient for clarifying thought, and that it will serve as a step towards some future picture of plant construction, which will achieve a higher degree of adequacy.

[1] Willis, J. C. (1902), pp. 420–1.
[2] For criticisms of the partial-shoot theories discussed by the present writer, see, on the leaf, Philipson, W. R. (1949), and, on the root, with an alternative interpretation, Allen, G. S. (1947).

CHAPTER IX

REPETITIVE BRANCHING AND THE *GESTALT* TYPE, WITH SPECIAL REFERENCE TO PARALLELISM

THE mammals, as the most highly organised of animals, in some respects occupy a zoological position corresponding to that of the angiosperms among plants. One of the sharpest differences in the contrasting structure of these two groups, is that the individuality of the mammalian body is of a much more fixed character; that body consists of a limited number of members and organs, which were already, once and for all, marked out in the embryo, and which have no power of subsequent self-multiplication. In the animal body, with its parts thus arranged in an ordered hierarchy, there is no such thing as an indefinite succession of limbs, and branches of limbs, numerically unfixed, and liable to impede one another; but this is what we find among plants, in which the urge to self-maintenance tends to the production of an indefinite number of growing points, and finds its expression in repetitive branching. A tree, or even a herb, as a rule consists of a whole series of shoot generations, which remain connected permanently with their progenitors and successors. As Goethe said long ago, the plant can be called a single individual only at the period when, in the seed condition, it has become detached from the mother plant.[1] Each lateral shoot may be regarded as a fresh unit which repeats the characters of the parent shoot; the whole plant is thus a matriarchal tribe of shoots, many generations of individuals living together, in dependence upon one another. In the upshot this leads inevitably to a certain competition between the generations, which, as we have already seen, profoundly influences their morphology.

This repetitive growth is characteristic, not only of the shoot as a whole, but also of the partial-shoot which forms the leaf, where

---

[1] Troll, W. (1926) [Goethe, J. W. von, *Vorträge*, 1796, III. *Ueber die Gesetze*, p. 356; see also *Zur Morphologie*, p. 117].

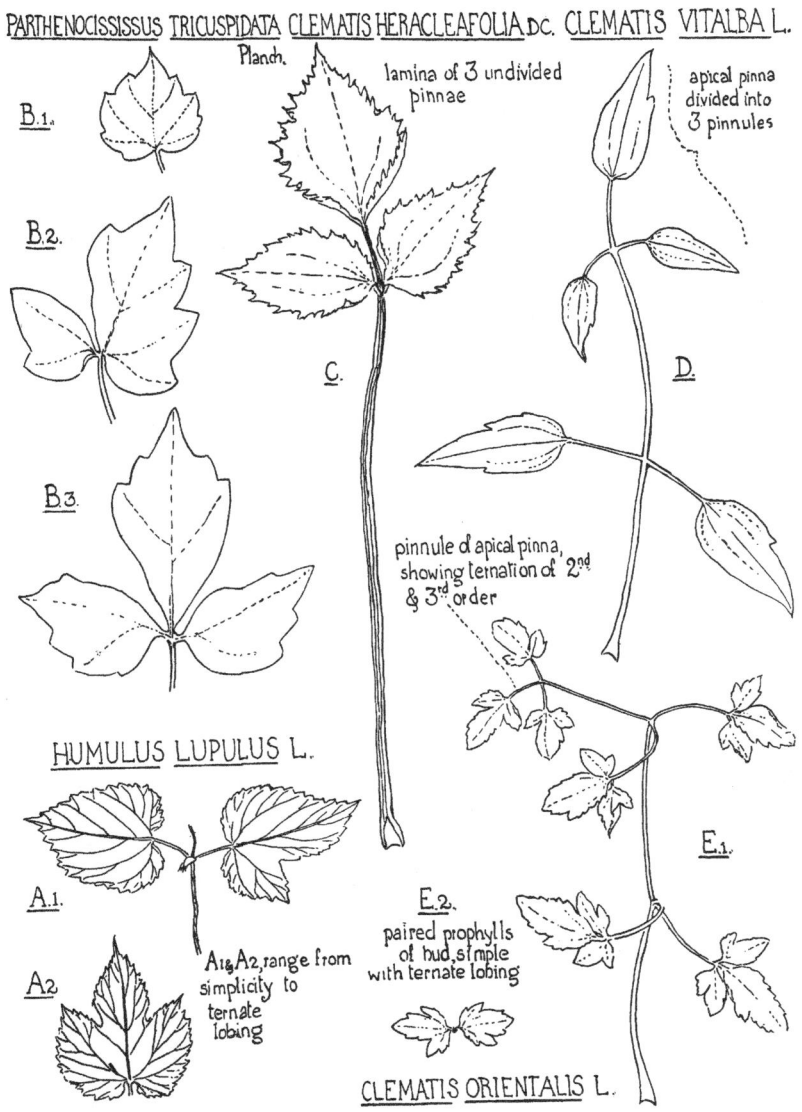

PARTHENOCISSISSUS TRICUSPIDATA CLEMATIS HERACLEAFOLIA DC. CLEMATIS VITALBA L.
Planch.

B.1.

lamina of 3 undivided
pinnae

apical pinna
divided into
3 pinnules

B.2.

C.

D.

B.3.

pinnule of apical pinna,
showing ternation of 2ⁿᵈ
& 3ʳᵈ order

HUMULUS LUPULUS L.

A.1.

E.1.

A₁&A₂ range from
simplicity to
ternate
lobing

A₂

E.2.
paired prophylls
of bud, simple
with ternate lobing

CLEMATIS ORIENTALIS L.

Fig. 24. Repetitive ternation in leaves. All drawings (×½).

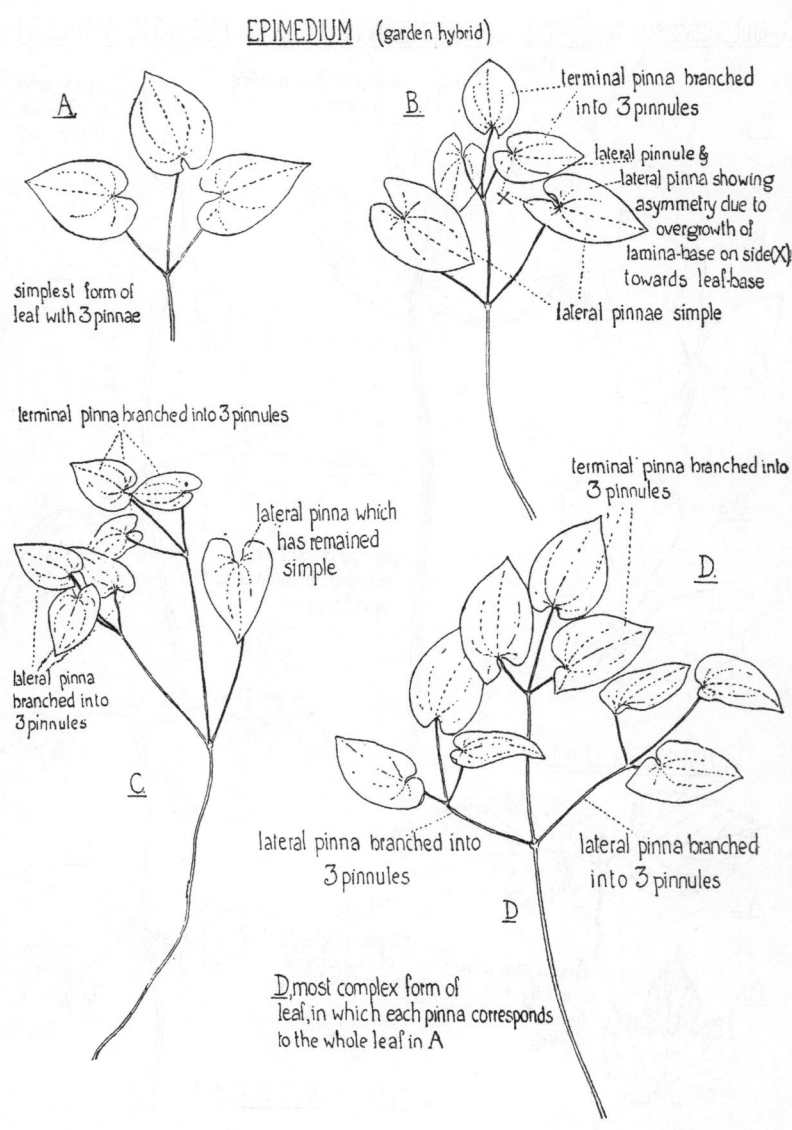

EPIMEDIUM (garden hybrid)

A.

simplest form of
leaf with 3 pinnae

B.

terminal pinna branched
into 3 pinnules

lateral pinnule &
lateral pinna showing
asymmetry due to
overgrowth of
lamina-base on side(X)
towards leaf-base

lateral pinnae simple

terminal pinna branched into 3 pinnules

lateral pinna which
has remained
simple

lateral pinna
branched into
3 pinnules

C.

D.

terminal pinna branched into
3 pinnules

lateral pinna branched into
3 pinnules

lateral pinna branched
into 3 pinnules

D.

D, most complex form of
leaf, in which each pinna corresponds
to the whole leaf in A

Fig. 25. Ternation in a whole lamina and its pinnae (× ⅓).

138

it results in a phyllome branched in various degrees. Innumerable cases might be cited in which a compound ternate leaf clearly reveals its equivalence to a simple leaf—that is to say, in which the two main lateral veins of the simple leaf have acquired the status of midribs, and have become associated with laminae of their own. The passage from a simple to a ternately lobed or divided leaf can be followed in the hop (*Humulus*, Fig. 24, A 1 and A 2, p. 137); *Clematis* (Fig. 24, C–E); and *Parthenocissus* (Fig. 24, B 1–B 3). This process may be repeated, the pinnae themselves suffering ternation, and even the pinnules so produced showing a further trifid division (*Clematis*, Fig. 24, E 1). In *Epimedium*, the whole leaf may be simply ternate (Fig. 25, A, p. 138), or one to three of the pinnae may again show ternation, so that each three-pinnuled pinna (Fig. 25, B–D) corresponds to the complete leaf in its simply ternate condition (Fig. 25, A). In pinnate leaves exactly the same thing happens. In *Jasminum humile* L.,[1] for instance, the leaf as a whole may have an undivided lamina, or it may consist of two, three, or more pinnae; moreover, these pinnae themselves may either be simple and undivided, or may consist of two, or of three, pinnules. There is thus a remarkably complete correspondence between shoot and leaf (Frontispiece, A). Another example, in which the equivalence of pinnae and pinnules is shown strikingly, is *Gleditschia caspica* Desf.,[1] in which the pinnae, even within a single leaf, may either be simple, or may consist of a large number of pinnules (Frontispiece, B). The same principle is illustrated in certain ferns, in which a pinna is as it were a copy of the whole frond, while a pinnule, correspondingly, may imitate a pinna.[2]

The equivalence, in descending order, which we have traced between a leaf, a pinna, and a pinnule, may be carried a stage further. Light is thrown on the subject by the surprising leaf forms of some members of *Eryngium*[3] (Umbelliferae). In *E. amethystinum* L., for instance, the pinnae in the terminal region of the leaf are twice branched, and bear marginal hairs. In passing down towards the leaf-base, the pinnae become progressively

---

[1] On *Jasminum* and *Gleditschia* cf. Leavitt, R. G. (1909), pp. 49–51.

[2] Leavitt, R. G. (1909), pp. 33–5.

[3] Urban, I. and Möbius, M. (1884); Möbius, M. (1884), (1886); Domin, K. (1909); Wolff, H. (1913); Chodat, R. (1920).

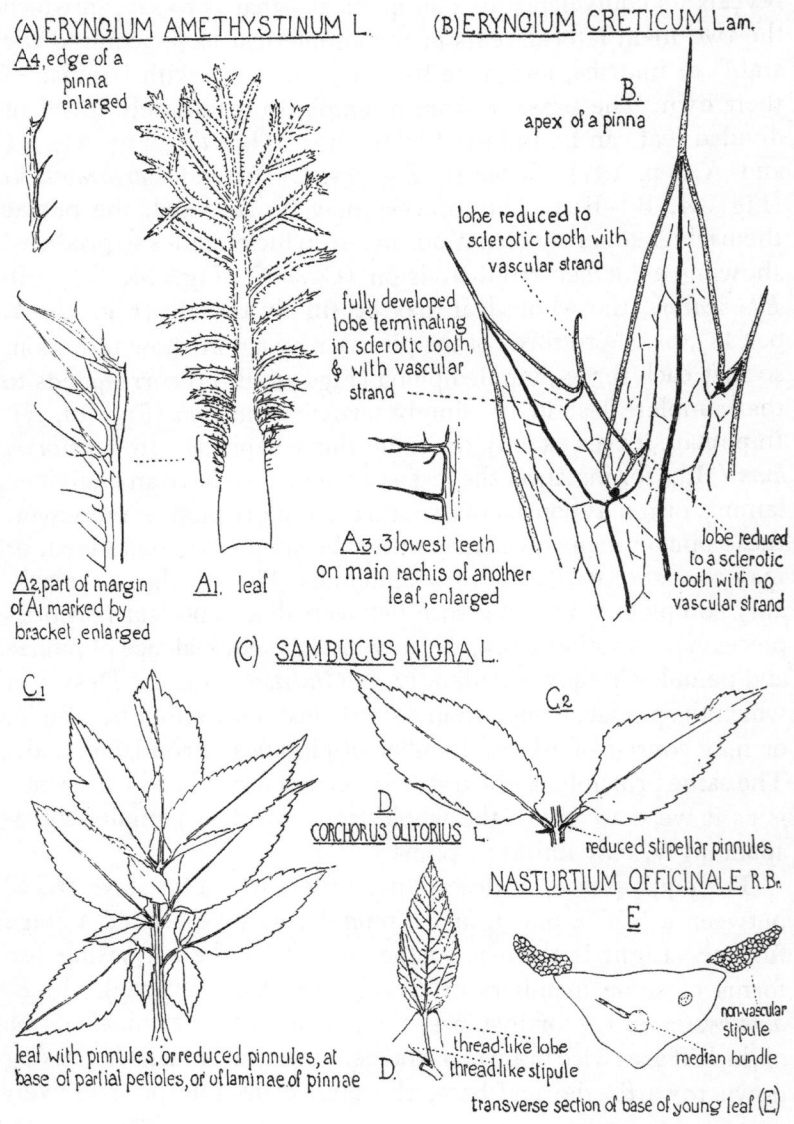

(A) ERYNGIUM AMETHYSTINUM L.

A4, edge of a
pinna
enlarged

A2, part of margin
of A1 marked by
bracket, enlarged

A1. leaf

A3, 3 lowest teeth
on main rachis of another
leaf, enlarged

(B) ERYNGIUM CRETICUM Lam.

B.
apex of a pinna

lobe reduced to
sclerotic tooth with
vascular strand

fully developed
lobe terminating
in sclerotic tooth,
& with vascular
strand

lobe reduced
to a sclerotic
tooth with no
vascular strand

(C) SAMBUCUS NIGRA L.

C1

C2

reduced stipellar pinnules

leaf with pinnules, or reduced pinnules, at
base of partial petioles, or of laminae of pinnae

D. CORCHORUS OLITORIUS L.

D. thread-like lobe
thread-like stipule

NASTURTIUM OFFICINALE R.Br.

E

non-vascular
stipule

median bundle

transverse section of base of young leaf (E)

Fig. 26. Transition from pinnae to hairs. A 1, C 1, C 2, D (×⅓); A 2–A 4
(enlarged); B (× 9 *circa*); E, very young leaf-base (× 31 *circa*).

140

simpler and smaller, until, by a perfectly graded transition, they pass into hairs. These changes are illustrated in Fig. 26, A 1–A 4, p. 140). The range between a pinna-lobe, and a mere sclerotic tooth without even a cell-lumen, can be followed in Fig. 26, B, which shows the apex of the pinna of another species, *E. creticum* Lam., rendered transparent by treatment with dilute potash, and then stained with gentian violet and eosin. In connexion with the relation of pinna-lobes and hairs, it may be recalled that Lund,[1] as long ago as 1872, came to the conclusion that the distinction between phyllomes and trichomes rests upon a quantitative rather than a qualitative basis.

An example of gradation between leaflets, and small pointed structures without laminae, may be found in the elder. Here the leaf is, as a rule, once pinnate, but sometimes, when the growth is luxuriant, certain pinnae bear lateral pinnules at the base; these may either be similar to the primary pinnae, though on a smaller scale (Fig. 26, C 1, p. 140), or they may be reduced to little horn-like bodies (Fig. 26, C 2).

Corchorus olitorius L. (Tiliaceae, Fig. 26, D) offers an instance of rather a different kind. Here the main part of the lamina is simple in outline, with a venation intermediate between pinnate and palmate, but, at the base on either side, there is a free lobe, almost hair-like in character, into which the outermost of the lateral veins enters. This hair-like form is repeated in the stipules. Among flowering plants in general, it has been shown that there is a tendency for exstipulate families to have entire laminae; this, again, points to a relation between the stipules of the leaf-base, and the teeth of the leaf-blade.[2]

Among themselves, stipules grade from members recalling fully developed foliage-leaves, down into hair-like bodies.[3] The extreme of reduction is seen in certain crucifers,[4] which are generally described as exstipulate, but which actually develop minute, non-vascular, paired structures at the base of the leaf, the stipular nature of which can scarcely be doubted (Fig. 26, E).

---

[1] On S. Lund's work see Foster, A. S. (1936 b).

[2] Sinnott, E. W. and Bailey, I. W. (1914); see p. 449.

[3] Cf. Glück, H. (1919), p. 42.

[4] Norman, J. M. (1857, 1858); cf. Arber, A. (1931 a), fig. 8, p. 186. See also p. 99 of the present book.

141

Possibly the glandular hairs of *Coriaria myrtifolia* L. (Fig. 8, B3–B6, p. 90) fall into the same category.

We have now traced a series of gradations, connecting large and complex foliage-leaves—in which many of the characteristics of shoot systems find their parallel—with small, non-vascular structures, which descriptive morphology would undoubtedly label as hairs. The conclusion is thus confirmed that every part of the above-ground, vegetative body, from a complex branch system, down to a mere hair, may be assigned to the shoot category. Long ago, Sachs,[1] with his flair for interpretation, approached this result, for his scrutiny of the facts inclined him to the paradoxical view that, in different flowering plants, the ovule might represent (in metamorphosed form) either a caulome; an entire leaf; a pinna; or a trichome. If the ovule can be equated with members of these four categories, the equivalence of these categories cannot but follow.

The relation to one another of a compound leaf, a simple leaf, and a mere lobe or hair, may perhaps be described as *identity-in-parallel*. A leaflet of a compound leaf comes in, as it were, on both sides of the equation: to the compound leaf, the leaflet stands in the relation of *part* to *whole*, but it is also the *equivalent* of the compound leaf *as a whole*, though in another generation (cf. Frontispiece).

When we consider the comparison of a leaf to a branching shoot system,[2] we see that there is a difference in the *degree* of union of different generations: in the leaf these generations are much more intimately connected than in the shoot. In correlation with this lack of independence within the leaf, the activity of the lateral growing points is reduced in each generation, and often, within a small number of generations, it sinks to nothing, so that growth ceases. In the shoot, on the other hand, as branch is developed from branch in comparative independence, though there is a diminuendo in the growth capacity of successive generations, this reduction is much slighter and slower. If, indeed, more independence for the branches of a shoot is secured artificially, as in propagation by cuttings, a series of lateral growing points may continue indefinitely in full vigour. It seems

---

[1] Sachs, J. von (1875), p. 428.

[2] On the comparison of leaf and shoot meristems, see Schüepp, O. (1938).

as if, by keeping its offspring in close union with itself, the leaf has thus sacrificed, for them and for itself, the potential immortality which the shoot retains.

From the earliest days we find hints in the literature that botanists have been intrigued by the problems of repetitive growth, at which we have been glancing. That Theophrastus had failed to arrive at consistent ideas about such questions is indicated by his distinguishing stem, branch, and twig, as three of the primary elements of the plant body;[1] he clearly realised, however, that both simple and compound leaves fall into one category, thus recognising the identity-in-parallel of a branch system with its own branches.[2] Jung, who was, as we have seen, a seventeenth-century representative of the classical tradition, was aware of the same truth; moreover he held that, when a plant had several 'crowns', the shoots from these crowns might be treated as each equivalent to an entire plant.[3]

Hitherto we have considered repetitive growth only as shown in the vegetative shoot, but botanists have always noticed the existence of this phenomenon in the reproductive stage. Theophrastus used the word καρπός both for a single fruit and also for a fruit cluster,[4] and he discusses how far a bunch of grapes or an ear of corn can be compared with a single apple fruit with numerous seeds.[5] He uses ἄνθος for a simple flower, such as rose or narcissus,[6] and also for the whole inflorescence of elder,[7] though recognising that the latter is made up of small individuals.

Other pre-modern writers were conscious of the close relationship between flowers and inflorescences, without, however, exactly defining it. Malpighi,[8] for instance, called the whole spadix of *Arum*, 'flos', while distinguishing the female flowers as 'flosculi'. Nehemiah Grew,[9] again, described the capitulum of the composites as a '*Flower*,' which itself "Embosomes, or is, a *Posy* of perfect *Flowers*". There is no need to multiply instances,

---

1 Hort, A. (1916), vol. I, pp. 10–11 [I. i. 4].
2 See Theophrastus' description of the sorb, p. 18.
3 Jung, J. (1747), cap. VI, 12, p. 14.
4 Hort, A. (1916), vol. I, pp. 240–1 [III. xii. 8].
5 Hort, A. (1916), vol. I, pp. 82–5 [I. xi. 4–6].
6 Hort, A. (1916), vol. I, pp. 90–1 [I. xiii. 2].
7 Hort, A. (1916), vol. I, pp. 246–7 [III. xiii. 6].
8 Malpighi, M. (1675), Pl. xxxi, fig. 184, and p. 49.
9 Grew, N. (1682), p. 170.

but, as one example from the nineteenth century, we may recall that Dresser[1] drew attention to the way in which a capitulum 'apes' a single flower, and, indeed, in our common speech, even to-day, a daisy is just a 'flower', like a buttercup.

It was not, however, until the present century that the resemblances between flower and inflorescence were subjected to a stringent analysis. This analysis has been the work of Wilhelm Troll,[2] whose detailed and comprehensive study of similarities, which most botanists either pigeon-hole as 'adaptations', or dismiss as idle curiosities, at least indicates that there is a wealth of significance implicit in such resemblances. How far he has succeeded in unearthing this significance is, however, another matter. In the upshot, Troll's work leads him to believe in the existence of what he distinguishes as the *Gestalttypus* of the flower; he contrasts this with the 'organisation type', which is the province of orthodox morphology. The meaning which Troll attaches to the term *Gestalt* will be best indicated by considering examples of the kind which he cites, but a preliminary analogy may make his rather unusual line of thought easier to follow. Suppose we imagine a group of art students, with a model posed before them, and suppose that one of them produces his impression of the figure in a charcoal outline, one as a water-colour, one in carved wood, one in a clay model, and so on. The results—since they are not produced in the same medium, and are thus built up out of different elements—cannot be said to belong to the same organisation type; but since they are all manifestations of the same model, they fall into the same *Gestalt* type.

When we use the term 'flowering plant', as synonymous with angiosperm, we are stressing the fact that, in the whole group, the reproductive phase assumes the particular form which Troll calls the flower *Gestalt*. This form may however appear—as we shall see in greater detail shortly—under the guise of a simple flower (euanthium); an inflorescence (pseudanthium); or an inflorescence of inflorescences. The 'flower', in the *Gestalt* sense, is thus not uniform in organisation, but may be constructed

---

[1] Dresser, C. (1859), p. 111.
[2] Troll, W. (1928*a*). Some account of Troll's work on the flower is given in pp. 179–82 of Arber, A. (1937*a*), which is partly incorporated in the present chapter.

in divers ways out of simple phyllomes or whole shoots. It is therefore impossible to give a precise verbal definition of the flower *Gestalt*, based upon the morphological nature of its constituents. In expressing his ideas Troll is hampered by the difficulty (which he recognises) of finding satisfactory terms—a difficulty akin to that which Goethe experienced in finding a name for his type appendage. It is clear that the words 'pseudanthium' and 'euanthium' convey a misleading impression, since, on Troll's view, the pseudanthium does not *simulate* the euanthium, but both, on the contrary, are manifestations of a single abstract type form, and one is no more 'pseudo' than the other. Furthermore, in treating Troll's ideas in English, we are faced by the difficulty that neither 'form' nor 'configuration' is adequate as a translation of *Gestalt*; it seems, indeed, best in many connexions to retain the German word, which has already found its way into our language in the expression 'Gestalt psychology'.

When we turn to Troll's detailed studies, we find that what, on a casual interpretation, might be called the 'mimicry' of the euanthium by the pseudanthium goes much further than is generally realised. Among the many remarkable similarities with which he deals, is that between the flower of *Aristolochia*, and the inflorescence (*Scheinblüte*) of *Cryptocoryne*. He also studied the surprising pseudanthia produced by the genus *Euphorbia*, which resemble simple flowers, even in unexpected minutiae. He compares, for instance, the involucre of *Euphorbia fulgens* Karw. (Fig. 27, A 1 and A 2, p. 146) with the corolla of *Nerium Oleander* L. They are alike in each having a basal tube with free distal lobes; in the presence of appendages at the mouth of the tube; and in the contorted aestivation of the free lobes. Another conspicuous example of a pseudanthium is offered by *Houttuynia cordata* Thunb. (Saururaceae, Fig. 27, B, p. 146). Here the inflorescence suggests a simple flower of one of the Ranunculaceae.[1] Like a flower in its perianth, the inflorescence-bud is enwrapped in its bracts. These bracts resemble a corolla, not only in the fact that they wither and fall, but also in their papillose surface (Fig. 27, B 4), which gives them exactly the aspect of white petals.

[1] Hutchinson, J. (1935).

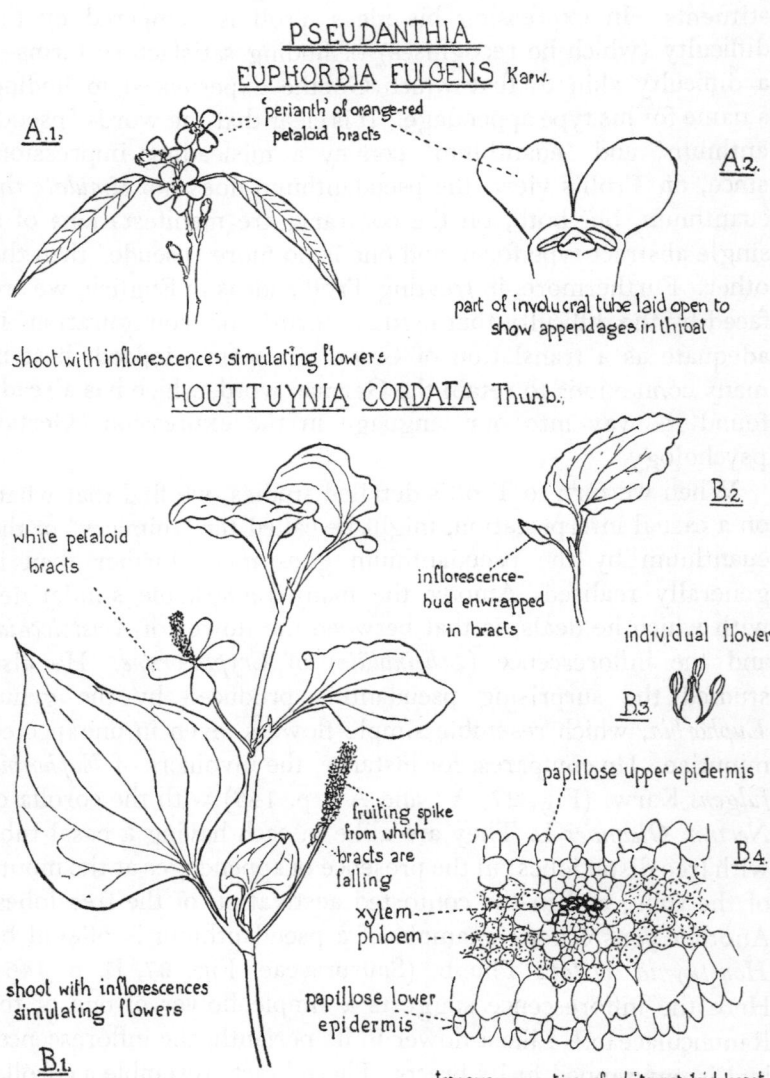

PSEUDANTHIA
EUPHORBIA FULGENS Karw.

A.1.

'perianth' of orange-red
petaloid bracts

A.2.

part of involucral tube laid open to
show appendages in throat

shoot with inflorescences simulating flowers

HOUTTUYNIA CORDATA Thunb.

B.2.

white petaloid
bracts

inflorescence-
bud enwrapped
in bracts

individual flower

B.3.

papillose upper epidermis

B.4.

fruiting spike
from which
bracts are
falling

xylem
phloem

shoot with inflorescences
simulating flowers

papillose lower
epidermis

B.1.

transverse section of white petaloid bract

Fig. 27. A1, B1, B2 (×½); A2, B3 (enlarged); B4 (× 193 *circa*).

146

It is the Compositae, however, that furnish the most obvious instances of pseudanthia, and it is from this family that most of Troll's material is drawn. The similarity of the involucre to a calyx, and of the ray florets to a corolla is especially noticeable in some composites in which the number of bracts and ray florets is small. *Chrysogonum virginianum* L., for example, has an involucre consisting of five bracts in one whorl, recalling a calyx, followed by five ray florets, suggesting petals—the whole producing a markedly flower-like effect.

The fact that Troll's emphasis is on form—as interpreted by the naked eye—rather than on organisation, leads him to the view that the *Gestaltlehre* of the flower should deal primarily with the perianth. Is it a certain puritanical strain in the scientific temperament that induces a reluctance to accept this conclusion? The prettiness of the petals, and their poetical associations, seem to remove them a little from the plane of biological thought, and botanists in general have tended to concentrate upon the 'essential organs', and have been content to relegate an interest in the perianth to the poet, the artist, and the gardener. Troll, on the contrary, has shown the importance, from the *Gestalt* standpoint, of even so apparently superficial a character as petal hue. He recalls, for instance, the old observation[1] that, alike in simple flowers, and in the capitula of the Compositae, where there is both blue and yellow coloration, the yellow is central and the blue peripheral (cf. *Myosotis, Solanum Dulcamara* L., various asters, etc.). Moreover, by a statistical analysis of the colour groups to which the petals of the Choripetalae, the petals of the Sympetalae, and the ray florets of the Compositae, can be assigned, he shows that the pseudopetals of the Compositae agree in this respect with the petals of the Choripetalae, rather than with those of the other members of their own grade, the Sympetalae. This agreement can be recognised, not merely in the ground colour of the ray florets, but also in the form and pattern of their colour markings (*Zeichnungen*). Troll shows that this 'brush-work', as it might be called, on the rays of the composite capitulum, resembles that found on the petals of the Choripetalae, rather than that on the lower lips of dorsiventral sympetalous flowers—the category

[1] Voigt, F. S. (1816), *Die Farben der organischen Körper*, Jena, pp. 57-9. The writer has not seen this book, which is quoted in Troll, W. (1928*a*), p. 112, footnote.

to which these rays actually belong, according to the canons of 'organisational' morphology. In the Compositae, the ray florets are not the only elements out of which a pseudocorolla may be constructed. In *Helipterum Manglesii* F. Muell., for instance, the yellow disk florets are surrounded by petaloid bracts, giving much the effect of a corolla. Moreover the capitulum of the Compositae may find its reiteration among plants of other families. For instance, in *Hacquetia Epipactis* DC., of the Umbelliferae, the inflorescence, with its yellow disk of flowers, surrounded by five or more yellow-green bracts, looks like the head of a composite.

The flower *Gestalt* may be expressed not only in a single flower, or in an inflorescence; it may even dominate the entire organism. There is a minute and curious Crassulaceous plant of South Africa, *Pagella Archeri* Schonl., in which the disk-shaped stem bears on its top a large number of closely packed sessile flowers, surrounded by two to three quasi-rows of crowded foliage-leaves. According to the original description,[1] the whole plant, which was often less than 2 cm. in diameter, bore a close resemblance to the flat capitulum of some Compositae.

So far we have been thinking only of a single inflorescence (or an inflorescence associated with foliage-leaves) which simulates a flower; but the complexity may go further, and an *inflorescence of inflorescences* may adopt the form of a single flower. *Syncephalantha decipiens* Bartl. (Compositae) has, as its generic name suggests, a reproductive shoot of this type. There is a terminal capitulum, around which, as a rule, five lateral capitula are grouped, all six standing almost at the same height. The central capitulum possesses the disk type of floret alone. The lateral capitula, on the other hand, though consisting partly of disk florets, each have also one or two ray florets, on the side remote from the central capitulum; the whole group is thus deceptively like a simple capitulum with disk and ray florets.

The reproductive shoot in a family so remote as the Gramineae offers a term of comparison for Troll's examples of compound capitula among the Composites. A grass panicle is a compound inflorescence in which each ultimate member is, usually, a simple inflorescence, the so-called spikelet; but there may be a still further

[1] Schonland, S. (1921).

degree of complication. In the bamboo genus, *Schizostachyum*,[1] it has been shown that the ultimate inflorescence branches are not mere spikelets, as was formerly supposed, but something more elaborate. In their younger stages they are bract-covered and externally spikelet-like, but they differ from true spikelets in structure and history, and they have been distinguished as pseudospikelets.[2] Each terminates in a spikelet, but is also capable of giving rise to a succession of spikelet-bearing branches from the axils of the bracts below the end spikelet. Each pseudospikelet may hence contain within itself a series of shoots of different orders. We thus see that the uniform spikelet *Gestalt* of the Gramineae can be achieved at different levels of organisation.

Apart from forms with pseudospikelets, another type of grass inflorescence in which complexity is concealed, finds its parallel among the Compositae. In the female plants of the genus *Spinifex* (Gramineae), the one-flowered spikelets, enclosed in the two outer empty glumes, are grouped into a spherical head. Such a composite as *Echinops globifer* Jacka, offers a comparable structure, since each of its capitula consists of one flower enclosed in a small involucre, these one-flowered capitula being grouped into secondary heads. The apparent simplicity of such inflorescences is due, in reality, to a sophisticated simplification.

The umbel of the Umbelliferae is another example of an inflorescence of inflorescences, and sometimes the equivalence of the partial and total inflorescences is indicated by the substitution of an umbel for an umbellule. Fig. 28, A, p. 150 shows a particularly vigorous head of *Daucus Carota* L., in which the bracts are replaced by foliage leaves, while the outer umbellules are represented by umbels, or even by a shoot terminating in an umbel, and bearing a second generation of umbels laterally. The Umbelliferae, indeed, offer specially favourable material for the study of the compound inflorescence as a form of the flower *Gestalt*. For instance, in the umbel of the hare's-ear (*Bupleurum rotundifolium* L.) drawn in Fig. 29, A1, p. 152, six of the um-

---

[1] McClure, F. A. (1934).

[2] It is possible that certain structures in the spikelets of bamboos, formerly interpreted as vegetative buds (e.g. Arber, A. (1934), fig. 35, D, p. 109) may be in reality the rachis buds of the pseudospikelets of McClure. Since Arber, A. (1934), was published, the writer has found corresponding structures in the spikelet of *Schizostachyum longispiculatum* Kurz, from the Buitenzorg Garden.

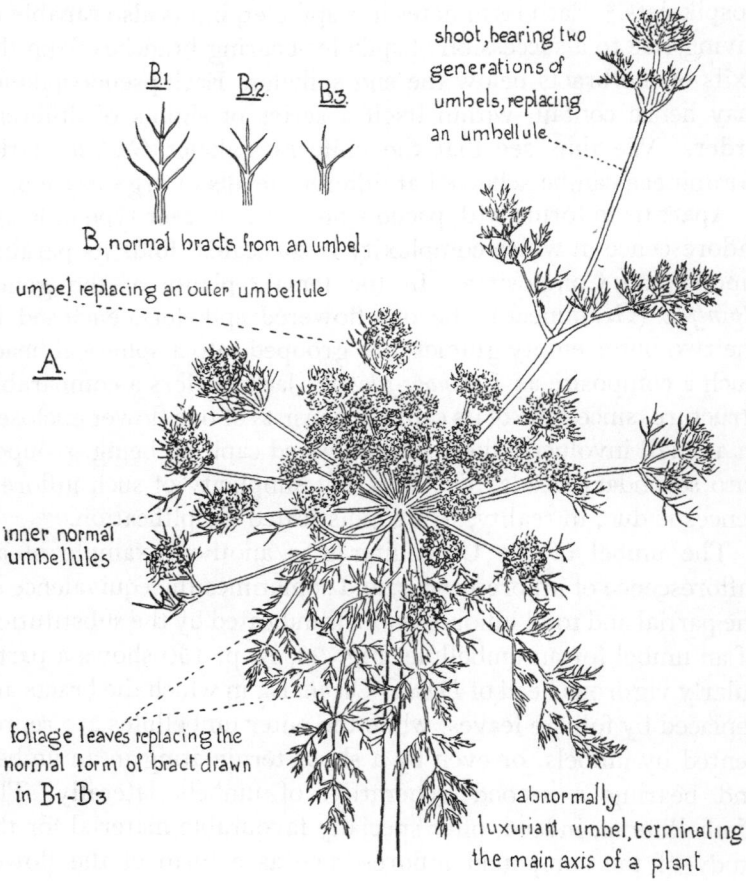

DAUCUS CAROTA L.

B_1. B_2. B_3.

B, normal bracts from an umbel.

shoot, bearing two
generations of
umbels, replacing
an umbellule.

umbel replacing an outer umbellule

A.

inner normal
umbellules

foliage leaves replacing the
normal form of bract drawn
in B_1-B_3

abnormally
luxuriant umbel, terminating
the main axis of a plant

Fig. 28. A, somewhat simplified; only 4 of 8 leafy bracts shown.
All drawings ($\times\frac{1}{2}$).

150

bellules each have three larger bracts towards the outside of the umbel, and one to four smaller bracts towards its centre. In the remaining umbellule, which is in the middle, there is no such sharp differentiation of bract size. The result is an aggregate inflorescence, in which the yellowish bracts of the partial inflorescences take the place of the outward extension of the ray floret corollas in the capitulum of the Compositae, or of the individual petals in the flower of one of the polypetalous dicotyledons. A corresponding and even closer resemblance can be traced in other umbellifers, such as *Coriandrum* (Fig. 29, B 1–B 2, p. 152), in which the place of the ray florets is taken by the outer flowers of each umbellule. These flowers show conspicuous asymmetry; the petals turned towards the centre of the umbellule are small, while those on its outer margin are expanded. In other genera the expanded petals may be confined to the region marginal to the umbel as a whole, as in *Daucus Carota* L. (Fig. 29, C). A like development may be seen in *Hydrangea*, amongst the Saxifragaceae, but here it is the sepals of the marginal flowers of the inflorescence that are disproportionately enlarged (Fig. 29, D).

Troll's purview is limited essentially to external form and coloration, but there are other resemblances between euanthia and pseudanthia which are worth a glance. It has been shown, for instance, that the behaviour of the residual vascular system in the extreme apex of certain inflorescences may be paralleled in the apex of a flower axis.[1] Moreover, some correspondence between inflorescences and flowers may be traced, not only in their structure, but in the order in which their parts come to maturity.[2] For instance, such a head as that of the teasle (*Dipsacus sylvestris* Mill.), in which opening of the flowers begins in the median region, and then works upwards and downwards, may be compared with an individual flower in which the succession of the parts is not strictly acropetal, the corolla making its appearance later than the stamens, despite its lower position on the flower axis.[3]

So far we have taken into consideration normal forms alone; but there are many abnormal transitions between flowers and

---

[1] Arber, A. (1940), pp. 624, 625.
[2] Suggestion by letter from Dr B. C. Sharman, 2 March 1939.
[3] Cf. Arber, A. (1937a), fig. 2, C, p. 164.

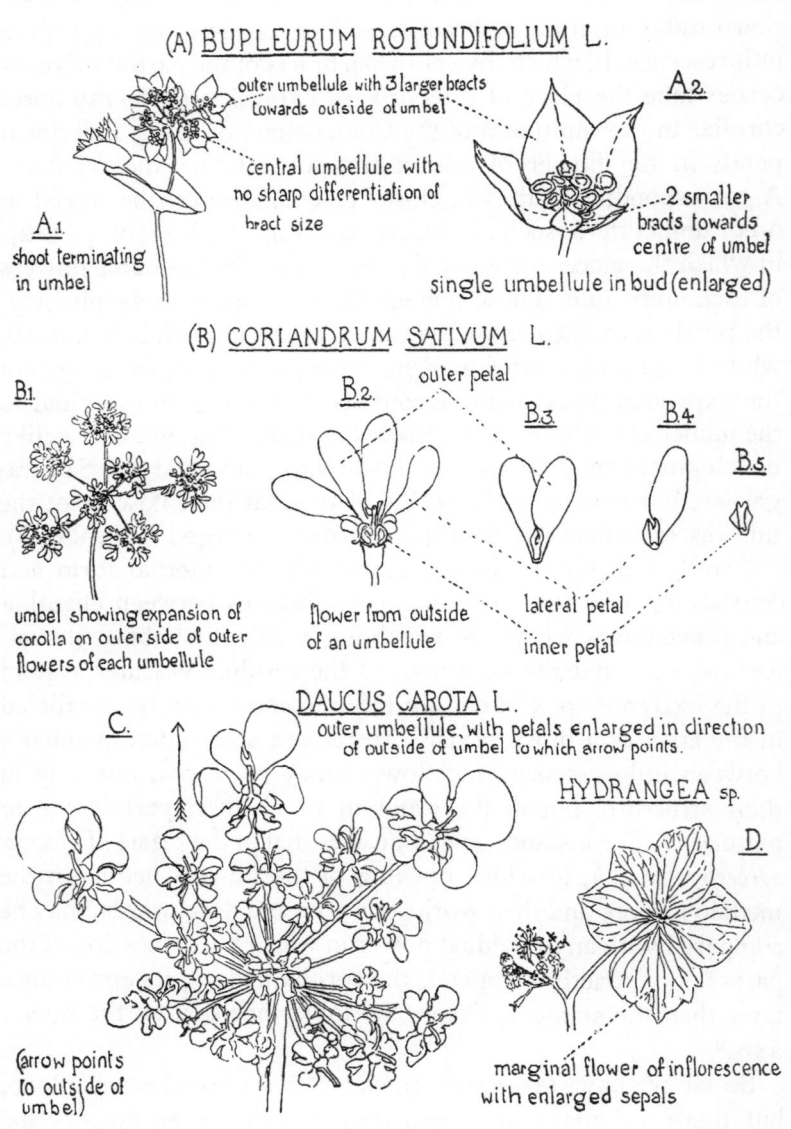

(A) <u>BUPLEURUM</u> <u>ROTUNDIFOLIUM</u> L.

outer umbellule with 3 larger bracts
towards outside of umbel

A_2

central umbellule with
no sharp differentiation of
bract size

A_1
shoot terminating
in umbel

-2 smaller
bracts towards
centre of umbel

single umbellule in bud (enlarged)

(B) <u>CORIANDRUM</u> <u>SATIVUM</u> L.

B_1

outer petal

B_2.

B_3.

B_4.

B_5.

umbel showing expansion of
corolla on outer side of outer
flowers of each umbellule

flower from outside
of an umbellule

lateral petal

inner petal

<u>DAUCUS</u> <u>CAROTA</u> L.

C.

outer umbellule, with petals enlarged in direction
of outside of umbel to which arrow points.

<u>HYDRANGEA</u> sp.

D

(arrow points
to outside of
umbel)

marginal flower of inflorescence
with enlarged sepals

Fig. 29. Transitions to pseudanthia. A 1, B 1, D (×$\frac{1}{2}$);
A 2, B 2–B 5, C (enlarged).

152

inflorescences, which suggest their equivalence. In the cultivated seakale (*Crambe maritima* L.), the floral axis may proliferate, and bear further flowers (Fig. 30, A 1 and A 2, p. 154). In *Plantago*—a genus notorious for deviations from its own norm—inflorescences may often be found to take the places of single flowers. In the spike of *P. Coronopus* L., sketched at the fruiting stage in Fig. 30, B, the bracts ranged from small simple structures of the normal type, to lobed foliage-leaves; in the axils of about thirty of them, there were sessile fruits (not visible in the drawing), while, in the axils of five others, there were stalked inflorescences. Among the Compositae it is not uncommon to meet with accessory capitula produced from the axils of the bracts of the primary capitulum; the 'hen-and-chicken' daisy (Fig. 39, D 1 and D 2, p. 179) is a variety in which this happens regularly, and the same thing has occurred in the garden marigold shown in Fig. 30, C. Sometimes such accessory heads may be, obviously, direct developments from florets, as in the abnormal capitulum of the goat's-beard (*Tragopogon pratensis* L.) drawn in Fig. 30, D 1, p. 154.[1] Here examples were found (e.g. Fig. 30, D 2, D 3), in which paired leafy structures, with a bud between them, took the place of the gynaeceum. This bud, on dissection, revealed itself as a miniature capitulum.

Troll not only elaborates the comparison between compound reproductive shoots (inflorescences) and simple reproductive shoots (flowers), but he follows it up by a perhaps rather fanciful suggestion as to the resemblance which a certain flowerlike grouping of parts, sometimes found *within* a simple euanthium, bears to the flower as a whole. In *Iris*, for instance, he distinguishes each of the three perianth members of the outer cycle, with its associated stamen and stylar branch, as forming a dorsiventral 'secondary flower'. This idea of the occurrence of a subflower *Gestalt* within a flower was foreshadowed long ago; the *Grete Herball* of 1526 says of *Crocus sativus* L. that "in the myddes of the floures sprynge thre chyves or small floures [the styles and stigmas] that be reed [red] which is saffron."[2]

Other writers before Troll had been feeling their way to the *Gestalt* concept. When Sachs wrote of "similarities which are

---

[1] Cf. Norman, J. M. (1857), footnote, p. 6; or (1858), footnote p. 115.
[2] Anon. (1526), *c*. ciii; cf. also Theophrastus on 'two-fold' flowers, p. 20.

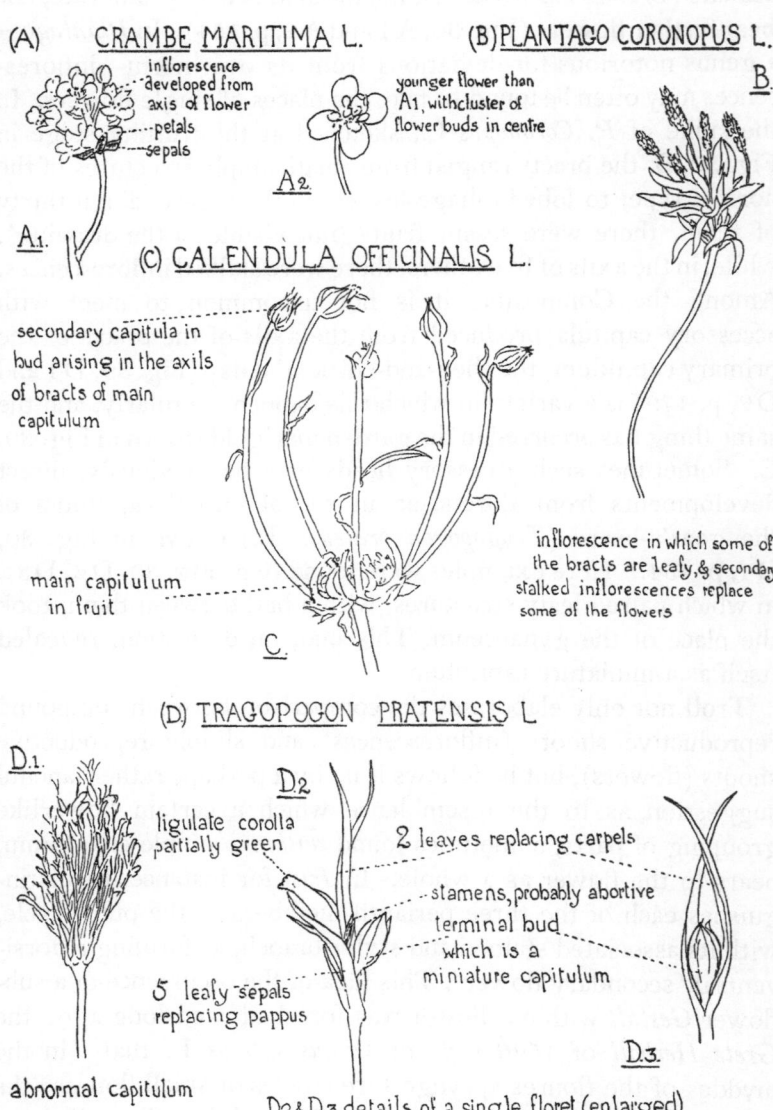

(A) CRAMBE MARITIMA L.

inflorescence developed from axis of flower
petals
sepals

A2.

younger flower than A1, with cluster of flower buds in centre

A1.

(B) PLANTAGO CORONOPUS L.

B.

(C) CALENDULA OFFICINALIS L.

secondary capitula in bud, arising in the axils of bracts of main capitulum

main capitulum in fruit

inflorescence in which some of the bracts are leafy, & secondary stalked inflorescences replace some of the flowers

C.

(D) TRAGOPOGON PRATENSIS L.

D.1.

D2

ligulate corolla partially green

5 leafy sepals replacing pappus

abnormal capitulum

2 leaves replacing carpels

stamens, probably abortive

terminal bud which is a miniature capitulum

D.3.

D2 & D3, details of a single floret (enlarged)

Fig 30. Inflorescences developed from floral axes.
All drawings (×½) except D2, D3 (enlarged).

154

intrinsically no similarities—well-defined forms, in which the form is, so to speak, accessory";[1] or when Macleod recognised what he called the "mechanical concordance" between the fruits of certain Myxomycetes and Gasteromycetes, and distinguished this concordance from homology;[2] these thinkers were approaching the *Gestalt* idea. Moreover, as Troll himself points out, this concept was, in some sense, anticipated by Delpino, in his work upon pollination mechanisms.[3] The chief difference between his outlook, and that of Troll, is that Delpino was essentially a teleologist, whereas Troll rejects adaptational explanations, and shows that the resemblances with which he deals are often carried into minute details to which no utility can reasonably be attributed. Allowing for this difference, Delpino's 'concetto'—which remains one, though founded on divergent schemes of organisation—may be identified with the *Gestalttypus*. As Delpino realised, the close similarity between, for instance, the flower of *Polygala* and of certain Papilionatae,[4] is developed upon the basis of a wholly different organisation; for the 'standard' in the Papilionatae, which is a single petal, is represented by two petaloid sepals in *Polygala*, and the 'keel', by one petal instead of two. Even in related plants, members of different whorls may simulate one another. There is, for example, much resemblance between the fruit head of *Trifolium procumbens* L., in which the characteristic form is due to persistent membranous *corollas* (Fig. 31, A 1–A 3, p. 156), and that of *T. fragiferum* L., with its correspondingly inflated membranous *calyces* (Fig. 31, B 1–B 3).

It has been held by Wettstein and others[5] that the simple flower in the angiosperms is to be interpreted as a reduced inflorescence, whereas it is more usually thought that the simple flower is the primary thing, from which the flowerlike inflorescence is derivative. Here, as elsewhere, the theory of shoot relations, which we are adopting, has the advantage of reconciling

[1] Pringsheim, E. G. (1932), p. 163 (a previously unpublished note by Sachs): "die organischen Formen bieten uns Ähnlichkeiten unter sich, die eigentlich keine Ähnlichkeiten sind, bestimmte Formen, bei denen die Form, sozusagen Nebensache ist".

[2] Macleod, J. (1919 and 1926), in both editions, pp. 79–82.

[3] Delpino, F. (1869).

[4] Cf. Troll, W. (1928 a), pp. 346 *et seq.*, and Delpino, F. (1869), p. 137.

[5] A full account of these theories will be found in Zimmermann, W. (1930), pp. 324 *et seq.*

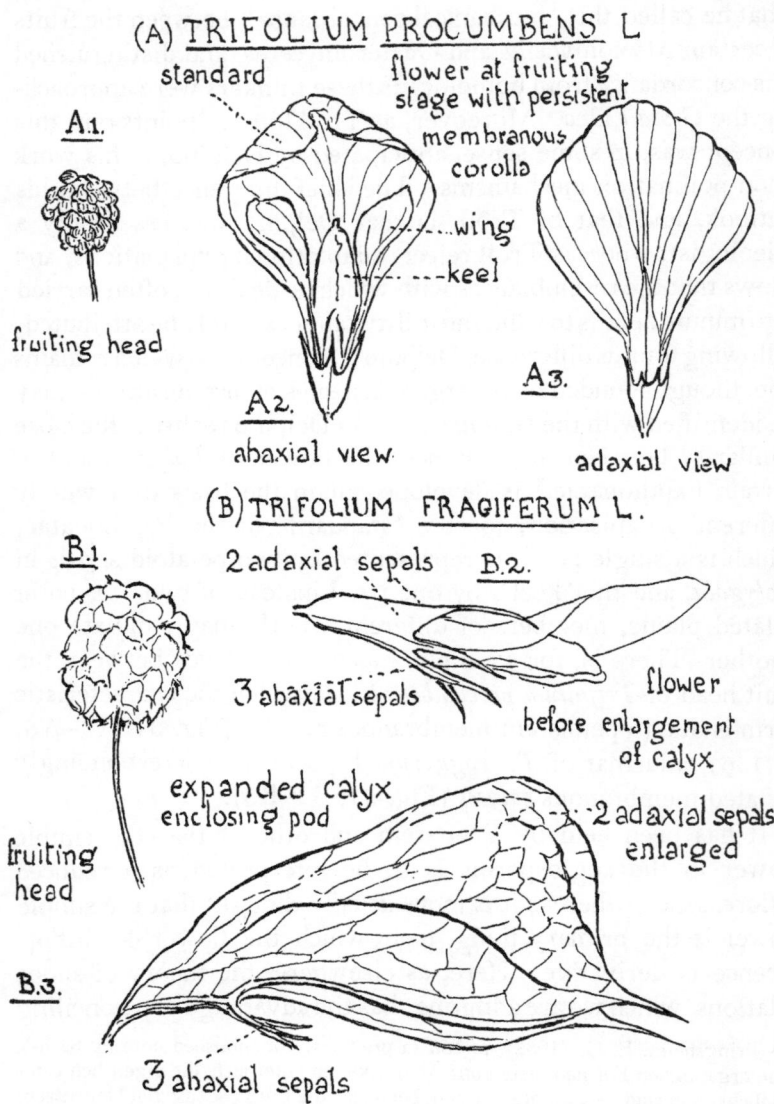

(A) TRIFOLIUM PROCUMBENS L

A.1.

standard

flower at fruiting
stage with persistent
membranous
corolla

wing

keel

fruiting head

A2.

abaxial view

A3.

adaxial view

(B) TRIFOLIUM FRAGIFERUM L.

B.1.

2 adaxial sepals

B.2.

3 abaxial sepals

flower
before enlargement
of calyx

expanded calyx
enclosing pod

2 adaxial sepals
enlarged

fruiting
head

B.3.

3 abaxial sepals

Fig. 31. Parallel development of corolla in *Trifolium procumbens* L. and calyx in *T. fragiferum* L. A 1, B 1 (nat. size); other drawings (enlarged).

views, which have been in the past opposed to one another; for, if we accept the idea that flower and inflorescence are both expressions of the flower *Gestalt*, though worked out in different shoot generations, the apparent antithesis between Wettstein's theory and more orthodox opinions finds its resolution. If we neither regard the pseudanthium as derived from the euanthium, nor vice versa, but consider the two forms as parallel and equivalent, the problem of derivation ceases to exist.

Though the most obvious examples can be found in external form, the *Gestalt* conception could equally well be deduced from the internal construction of the flower. The embryo-sac of angiosperms, for instance, with its characteristic 8-nucleate scheme, may be the equivalent of one, two, or four spores;[1] that is to say, the typical embryo-sac *Gestalt* type may be built up in several different ways.

So far we have concentrated attention upon reproductive shoots, but there is no difficulty in finding examples of *Gestalt* phenomena in the vegetative regions. There are, for instance, close similarities between certain leaves which are, however, alien to one another in organisation, such as the various peculiar phyllomes of the pitcher plants; if we compare the lids in the different genera, we see that that of *Nepenthes* is an abaxial member, while that of *Cephalotus* is adaxial. Nevertheless, the pitchers with their lids approximate in form in these two unrelated plants, though the means by which the result has been attained are essentially different.[2]

The analysis of the plant body into shoot complexes, simple shoots, and shoots of varying degrees of partialness—which we have been considering in this and the preceding chapters—raises the question of the exact relation of these elements to the plant as a whole. It is obvious that this relation differs fundamentally from, for example, the relation of groups of tessarae, individual tessarae, or fragments of tessarae, to a mosaic pavement formed out of them, considered as a whole. It may be suggested that such relations should be distinguished by calling each tessera-complex, or each individual or fragmentary tessera, a *component*[3]

---

[1] See Maheshwari, P. (1937), (1941), and (1948).   [2] Arber, A. (1941c).
[3] The word *section* is used, as a contrasting term to *part*, in an illuminating discussion in Hallett, H. F. (1930); but we are here substituting the term

of the whole mosaic, while a shoot, or leaf, would be, on the other hand, a *part*[1] of the whole plant. A *component* would then be one element, which, by merely additive association with other components, forms the whole, while a *part* is an advance upon a *component* in the fact that it not only acts the role of a component, but also, within its own boundaries, reproduces the whole. There is thus revealed, as Plotinus wrote, "the whole in all, and in every part the whole".[2] We have already cited[3] the classical distinction between the so-called *dissimilar parts* in the organism, such as a hand or a leaf, which are denatured if dissected, and *similar* parts, such as bone or wood, which still retain their 'bone-hood' or 'wood-hood', even if repeatedly subdivided; here we have in essence the distinction between parts and components.

The idea of a part of a plant, such as a lateral shoot or leaf, forming, in its own degree, a miniature representation of the whole plant, is, indeed, no new thing in botany. Early in the nineteenth century, Oken, whose extravagant nature philosophy is illumined here and there by flashes of genuine insight, wrote that, "A leaf is a whole plant."[4] Moreover the relationship, within the plant, of whole and part, recalls a certain conception of the relation of man to the universe, which, though occurring in classical literature, did not reach its fullest expression until the Middle Ages.[5] According to this conception, man, the *microcosm*, was regarded *both* as one of the innumerable constituents, which together make up the universe, and *also* as a mirror or symbol of that universe, seen, as it were, through a diminishing glass. In other words, man offers, within the limitations of his finiteness, a parallel to the total universe. So, comparing small things with great, we may call a shoot a

component, because, to biologists, the term *section* has already more than one special technical significance.

[1] *Parts*, in this sense, have been distinguished as *real parts*, or *differentiates*, or *parts-of-a-whole*; see Russell, E. S. (1933), p. 157; Stace, interpreting Hegel, uses *parts*, not in the sense of Russell's *parts-of-a-whole*, but of Hallett's *sections* (Stace, W. T. (1924), pp. 204–5).

[2] Bréhier, E. (1927), IV. 2. 1, p. 9: ὅλη ἐν πᾶσι καὶ ἐν ὁτῳοῦν αὐτῶν ὅλη.

[3] See pp. 10, 11.

[4] Oken, L. (1810), vol. II, p. 72, §1144: "Ein Blatt ist eine ganze Pflanze mit allen Systemen und Formationen."

[5] Arber, A. (1946b), pp. 224–8.

microcosm of the whole plant: while a leaf; a leaflet; a leaf-lobe; or even the hair to which a lobe may ultimately be reduced; is a microcosm of the whole shoot.

We saw at an earlier stage[1] that the partial-shoot theory of the leaf involved the merging of the concept of the *organisation type* in the broader concept of *parallelism* and thus the restatement of the Goethean theory of the *type appendage* in terms of *parallelism in organisation* within the plant body. If we apply the idea of identity-in-parallel to the *Gestalt* type also, we see that this concept may be replaced, correspondingly, by the idea of *parallelism in configuration* between phases of the plant body, which are not identical when viewed in their organisation aspect. The change from the type concept to that of parallelism has the advantage of detaching morphological thought from the tyrannic notion of irreversible historic derivation.[2] When organisms (or organs), viewed under the type concept, can be arranged in a series according to their degree of approximation to the type, this succession has often been treated as being actually, as a matter of history, derivative. When, on the other hand, such related forms are seen from the standpoint of parallelism, there is no question of a basic type to which they all conform; and if they can be arranged in a series, this series is recognised as an ordination of the mind, and thus may be read in either direction, without assigning primitiveness to one end, and derivativeness to the other.

When, though recognising the undoubted value of the type concept, we scrutinise it critically, the idea suggests itself that its origin is by no means wholly rational. It may perhaps be hazarded that its source—like that of Plato's 'forms'—is to be sought in the well-nigh irresistible desire to discern, in this mutable world, something changeless and eternal in which the mind can rest. That, to its more devoted adherents, the type concept contains an emotional element, is suggested by Troll's lament over Goebel's rejection of typology as 'almost tragic'.[3] If the idea of the type is, in truth, tinged with escapism and wishful thinking, this is a further justification for replacing it, where possible, by the more realistic notion of parallelism. This

---

[1] p. 86. [2] Cf. pp. 63 *et seq.*
[3] Troll, W. (1935, etc.), Vol. I, p. 31, footnote [1. 1].

replacement involves a change of outlook recalling that which has already taken place in the general study of botanical systematics. Here the importance of parallelism has long been recognised. As early as 1807, Corréa da Serra,[1] discussing the fruits of dicotyledons and monocotyledons, pointed out that their forms "se répétent dans les deux séries", and he went on to indicate that, though fruits paralleling those of certain dicotyledons had not, in his time, been found among monocotyledons, future discoveries might fill those blanks. John Stuart Mill,[2] again, in 1843, detected "parallelism in the ...natural orders of plants and animals", but he ruled it out as "an anomaly and an exception in nature"; he thus—with the self-protective instinct which so often hampers thought—averted his mind from a disturbing truth, which he had glimpsed and might have pursued. Later in the nineteenth century, when Darwinism was dominant, the idea of parallel seriation was little stressed, since it conflicted with the deification of chance. Darwin himself recognised that congeneric species might be expected occasionally to "vary in an analogous manner",[3] but he does not seem to have attached importance to this point. T. H. Huxley[4] was more alive to its significance; he held that genera and larger groups might arise polyphyletically, a process which could not but involve at least a secondary parallelism. Julius von Sachs[5] read the earliest copy of *The Origin of Species* to reach Germany, and at first accepted its teaching; but he grew steadily more and more critical of Darwinism, and eventually came to lay much stress upon parallelism in evolution. He believed, for instance, that heterospory, seed formation, and angiospermy, had arisen independently in different lines, and that monocotyledons and dicotyledons were essentially parallel rather than related groups (*unter sich kaum verwandt*). Other later writers have worked out the subject of parallel seriation in full detail, and the large part which it plays is to-day generally recognised.[6]

We have now considered the *type* concept—including in this term both organisation and *Gestalt* types—and we have seen that

---

[1] Corréa da Serra, J. F. (1807).   [2] Mill, J. S. (1843), vol. II, p. 135.
[3] Darwin, C. (1859), p. 161.   [4] Huxley, T. H. (1888), p. 123.
[5] Pringsheim, E. G. (1932).
[6] For a discussion with references see Arber, A. (1925), pp. 223–32; and on parallelism in the Gramineae, Arber, A. (1934), chap. XVII, pp. 380–409.

this concept may perhaps be absorbed into that of *parallelism*. We have not, however, tried to analyse the exact relation of these two ideas, each of which has undoubtedly a certain validity from its own standpoint. The type concept offers us a way of realising unity in the multiplicity of phenomena; in existing organisms and organs we can see, made variously explicit, potentialities which, in the type, are implicit. We are then faced with the old and ever-recurrent problem of the antithesis between the One and the Many; how are we to pass in thought from the type, which is unity in abstraction, to the actualised multiplicity of phenomena? Can we find a hint at an answer in the notion that the process by which the individual phenomena are reached from the type is that of parallel 'becoming'—this process being understood, not in the historic, but in the logical sense? Making unorthodox use of the terms of the Hegelian triad,[1] it might be said that, if *unity of type* is taken as the Thesis; the *multiplicity of individual forms* which, though distinct, all conform to the type, will be the Antithesis; and that this static Thesis and static Antithesis will be synthesised in the third term of the triad, '*parallel becoming*', which, in its dynamic character, emerges from and relates them both.

[1] Cf. Stace, W. T. (1924), pp. 92, 93, § 126.

CHAPTER X

THE MECHANISM OF PLANT MORPHOLOGY

HITHERTO we have discussed morphology from the comparative standpoint; it remains to consider how far plant form can be interpreted in terms of cause and effect, as these words are generally understood—that is to say, in terms of such 'causes' as physical and chemical factors. This aspect of the subject, which has been called 'causal morphology', has sometimes taken within its purview, not only physico-chemical factors, but also developmental history. One of the most impressive achievements in botany, during the last hundred years, has been the accumulation of data bearing on the mode of ontogeny from the fertilised egg, and also the development of the shoot from the meristematic apex. [1] The most recent work[2] tends to show that these two phases in the developmental history cannot be isolated from one another, since the organisation of the mature shoot is merely a modification of the original 'pattern' in the relevant part of the embryo. This holistic conception of plant ontogeny is irreconcilable with a claim that has sometimes been made, or implied, that a 'causal' explanation of plant form is to be sought in the developmental history. This claim involves a conception of development as consisting of stages which are discrete entities, each stage being treated as the 'cause' of that which follows it, and as being itself 'caused' by that which precedes it; but this is to confuse the succession in time of a continuous series, with causation. Two phases may follow one another "as the night the day", but this does not entitle us to call the first the 'cause' of the second. It is necessary for the biologist to achieve the difficult task of realising the individual organism as consisting of a unifi-

---

[1] A summary of the literature with a full bibliography will be found in Sifton, H. B. (1944); among modern work, that of Foster and his school on the growing apex is of primary importance (cf. Foster, A. S. (1939), etc.; on vascular development see papers by Esau, e.g. Esau, K. (1943).

[2] See especially pt. III of Miller, H. A. and Wetmore, R. H. (1945, 1946).

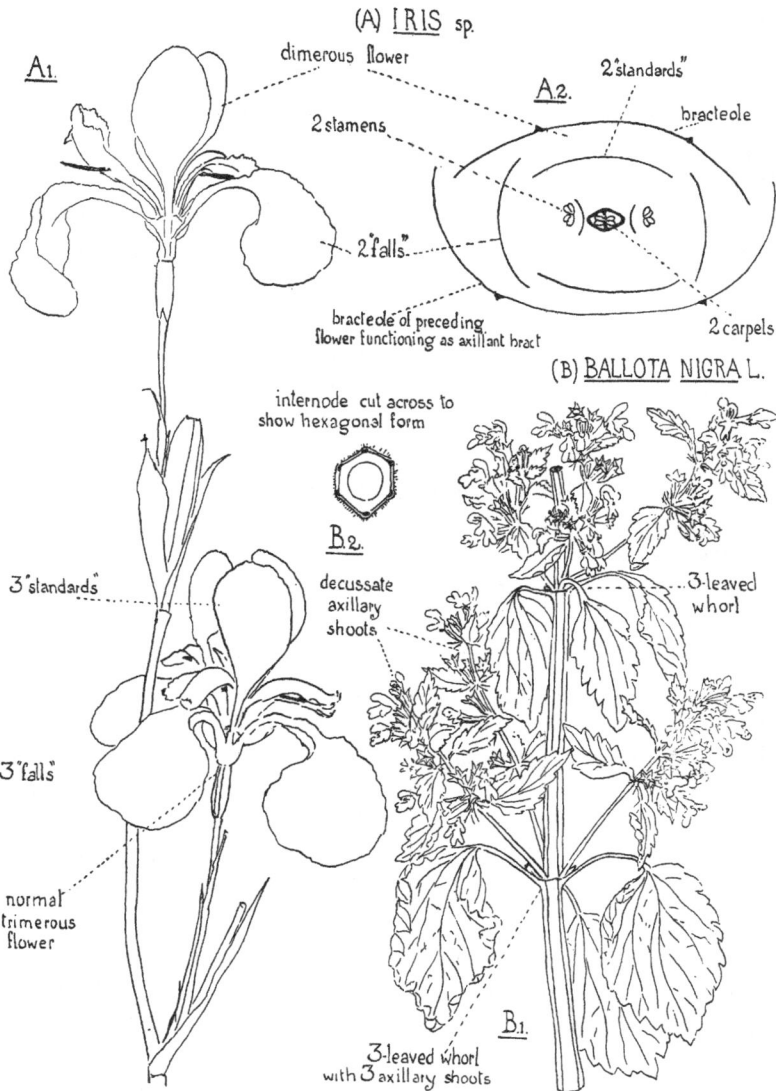

(A) IRIS sp.

A_1.

dimerous flower

2 stamens

2 "falls"

A_2.

2 "standards"

bracteole

bracteole of preceding
flower functioning as axillant bract

2 carpels

(B) BALLOTA NIGRA L.

internode cut across to
show hexagonal form

B_2.

3 "standards"

decussate
axillary
shoots

3-leaved
whorl

3 "falls"

normal
trimerous
flower

B_1.

3-leaved whorl
with 3 axillary shoots

Fig. 32. Changes in numerical rhythm. A 1, B 1 ($\times \frac{1}{2}$); B 2 (enlarged).

cation of every phase of its existence from the fertilised ovum onwards; it is a fatal error to see the individual as a summation of stages. These stages are not realities; they are the imaginary result of cutting a continuous series into segments, and then stringing them out in time. The human mind goes through this process for its own convenience, and then—forgetting the artificiality of the situation which it has created—it proceeds to endow its own constructions with a causal relation to one another.

If, then, we set aside the hope of finding the *causes* of plant form in developmental history, we may yet detect therein indications as to the *manner* in which some of the features of this form have come into being. It is clear, for instance, that phyllotaxis, and the numerical relations of the parts of the flower, depend upon the rhythmic development of primordia at the growing apex. A curious feature is that this growing apex may sometimes assume a rhythm different from that characteristic for the species, but yet this modification, though a departure from the normal, may be so consistent and regular that it does not suggest an abnormality, in the common acceptance of this term. It is not rare, for instance, to find a shoot of elder (*Sambucus nigra* L.), in which the growing apex must have put forth its primordia on six orthostichies instead of four, so that there are three leaves symmetrically placed at each node. Fig. 32, B, p. 163, shows a corresponding occurrence in a flowering shoot of black horehound (*Ballota nigra* L.), in which the usually four-square stem is hexagonal, and there are alternating three-leaved whorls, though the lateral branches, both vegetative and fertile, have all retained their binary symmetry. That the tetragonal arrangement in madder (*Rubia peregrina* L.) might vary with perfect neatness to a hexagonal scheme, was noted by Sharrock in the seventeenth century.[1] The apex of a flower may also undergo changes in rhythm, like those of a vegetative apex. Fig. 32, A, p. 163, shows a completely dimerous flower of *Iris*, on a shoot which also bore a flower with normal trimery, while in Fig. 33, p. 165, are sketched flowers of *Potentilla*,[2] with calyx and corolla regularly formed, but with their parts in three, four, five, or six. In these

[1] Sharrock, R. (1660), p. 145.
[2] On numerical variation and sectorial reduction in *Potentilla*, see Tansley, A. G. (1948).

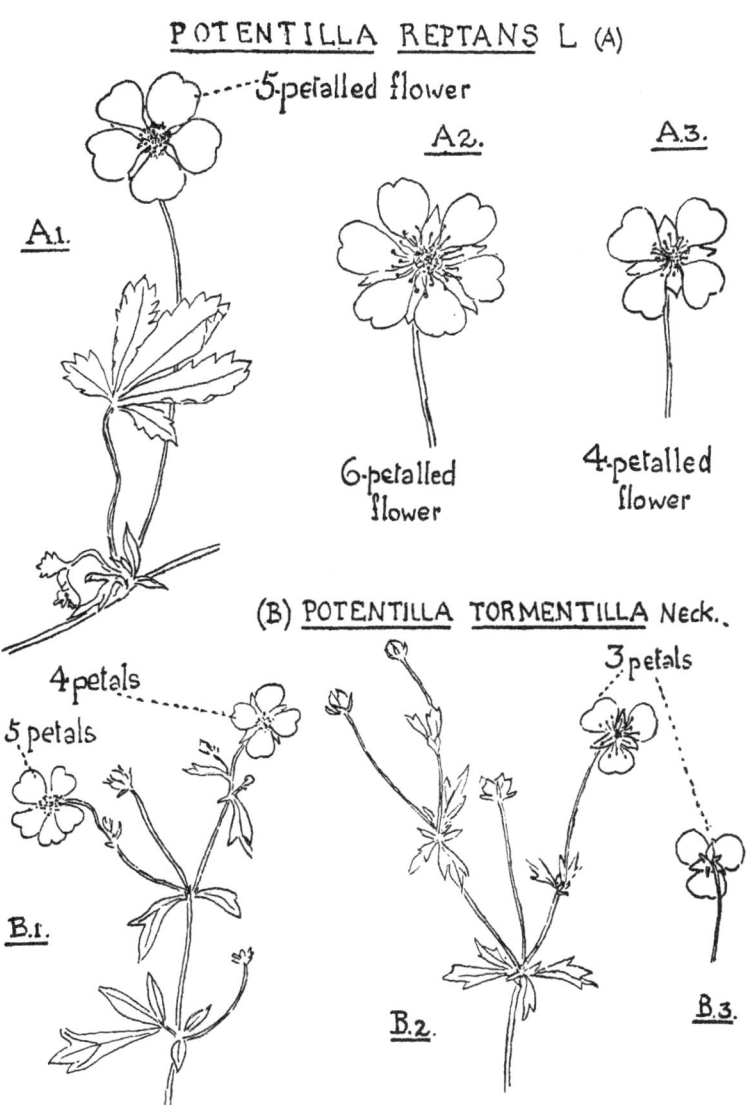

POTENTILLA REPTANS L (A)

A1.

-----5-petalled flower

A2.

A3.

6-petalled
flower

4-petalled
flower

(B) POTENTILLA TORMENTILLA Neck.

4 petals

5 petals

3 petals

B.1.

B.2.

B.3.

Fig. 33. Changes in numerical rhythm in *Potentilla*.
All drawings (nat. size).

numerical examples, the growing apex switches over bodily, as it were, to a new symmetry; but there are other cases in which some change affects a segment only of the shoot tip. In the red campion (*Lychnis diurna* Sibth.) drawn in Fig. 34, A 1, p. 167, a longitudinal segment of the stalk of the inflorescence was pale green, and the flower arising from this segment was white instead of pink. Such a change may, moreover, occur in the growing point of an individual flower, as in the example shown in Fig. 34, A 2, in which four petals were pink and one was white. The failure of anthocyanin development in this one petal was associated with paleness in the corresponding strip of calyx tissue, and its continuation down the pedicel; this strip must have originated from the same sector of the growing apex as the white petal. Chlorosis, the distribution of which follows similar lines, is seen in the leaf of meadowsweet in Fig. 34, B. Here all the large and small leaflets on one side of the rachis are white, while those on the other side are green; the chlorophyll failure must thus have been confined to a sector of the primordium. These examples are pathological, but, in normal flowers also, different sectors often show a difference in development.[1] That the sectors of a shoot possess an individuality of their own, has indeed long been recognised. De Candolle, for instance, wrote of a branch bearing whorls of leaves or floral parts as "comme composé de plusieurs fragmens soudés longitudinalement".[2] Such an analysis of the shoot finds its twentieth-century counterpart in Priestley's idea of the "unit of shoot growth", which he describes as "the segment of the axis which subtends a leaf initial and surrounds its leaf-trace as it differentiates".[3] This view is a modern development from various earlier theories which have treated, as a fundamental unit, a segment of the plant body, which some authors call a 'phyton',[4] consisting of an internode, its upper leaf, and sometimes a root. The whole corpus of phyton theories represents a search after *physiological* units, and thus belongs to a different level from the theory that the shoot (including the

[1] Goebel, K. von (1933), Ed. 3, pt. iii, p. 1853.
[2] Candolle, A. P. de (1827), vol. i, p. 532.
[3] Priestley, J. H., Scott, L. I., and Gillett, E. C. (1935).
[4] For an early adumbration of the phyton theory, see Spencer, H. (1867), vol. ii, p. 69; for more modern references see Wetmore, R. H. (1943), p. 16; Arber, A. (1930), p. 299; and, especially, Majumdar, G. P. (1947).

various forms of partial-shoot in this term) is the primary *morpho-logical* unit. In the growth-unit theories and the shoot theory, the same facts are interpreted from two independent standpoints;

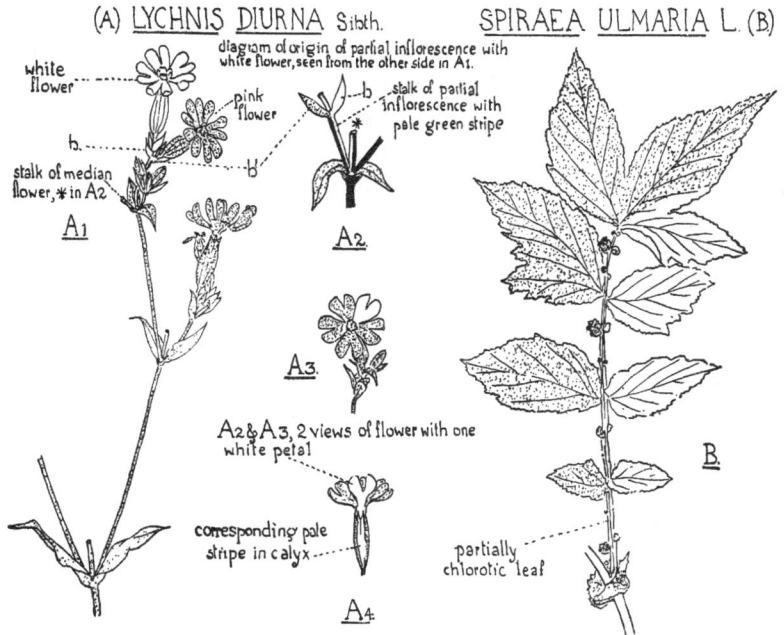

Fig. 34. Chemical characters confined to one sector of a growing apex. A, *Lychnis diurna* Sibth. A1, A2, male form (near Wenlock Edge, Shropshire, 4 July 1938): A1, shoot showing one white flower ($\times \frac{1}{2}$); A2, diagrammatic sketch of the origin of the partial inflorescence with its white flower; the flowers, being staminate only, eventually break off, leaving a segment of the pedicel as at * in A1 and A2; A3 and A4, a flower with one white petal ($\times \frac{1}{2}$) from a plant in which all the other flowers were pink (near Prospidnick, Helston, 5 Sept. 1938). B, *Spiraea Ulmaria* L., partially chlorotic leaf (near Wenlock Edge, 4 July 1938) ($\times \frac{1}{2}$).

these two types of theory can thus coexist on different planes of thought without conflicting.

When we consider the way in which plant form, as we see it at maturity, has been developed from its embryonic initials, we find that the main factor in this process has been 'relative' or 'differential' growth. This feature was hinted at by Herbert

X. THE MECHANISM OF PLANT MORPHOLOGY

Spencer[1] in 1867; fully discussed by D'Arcy Thompson[2] about half a century later; and, more recently, elaborated in modern and analytic terms by Julian Huxley.[3] The last writer treats it as "general growth which is quantitatively different in the three planes of space, or growth localized at certain circumscribed spots". Detailed research work on the principle of relative growth consists in elaborating and expressing in mathematical language certain facts with which we are all familiar. We recognise the faces of our acquaintances chiefly by differences in proportion, and such 'deformations' of a type—to use D'Arcy Thompson's term—can only have become manifest through differential growth, which may be analysed by the "method of coordinates".[4] As a botanical example we may take the fruits of the crucifers, which present a wide range of forms, though all these forms are reducible to a single scheme of organisation; the features which make them superficially so unlike are due to growth-emphasis taking contrasting directions. The gynaeceum in this family may be flattened either in the plane at right angles to the replum, or in that parallel to the replum, which then becomes extended like the valves. In angiosperms in general, any part of the shoot may, as a result of differential growth, become fantastically elongated in comparison with the norm. We will only recall the leaf of *Eryngium pandanifolium* Cham. et Schlecht.,[5] which may be nearly five feet long, but only 1½ inches broad at the base; i.e. about forty times as long as it is wide; the lengthy flower-tube of *Mirabilis longiflora* L., and the long ovary of *Glaucium leiocarpum* Boiss. Some such peculiar elongation may differentiate one related plant from another; for instance, the beak of the fruit in *Anthriscus*, below the styles, is minimal (Fig. 35, A, p. 169), while in the allied *Scandix* it may be conspicuously long (Fig. 35, B). Correspondingly, the beak of *Ranunculus falcatus* L. (Fig. 35, D) is greatly extended in comparison with the equivalent part in most members of the genus, e.g. *R. bulbosus* L.,

[1] Spencer, H. (1867), vol. II, p. 7.
[2] Thompson, D'Arcy W. (1917) (see also 2nd ed. (1942)).
[3] Huxley, J. S. (1932).
[4] Thompson, D'Arcy W. (1942), chap. 17, pp. 1026–95.
[5] Chamisso, A. de and Schlechtendal, D. F. L. von (1826), p. 236. (Pages misnumbered in the only copy available to the writer; this is probably 336 in other copies.)

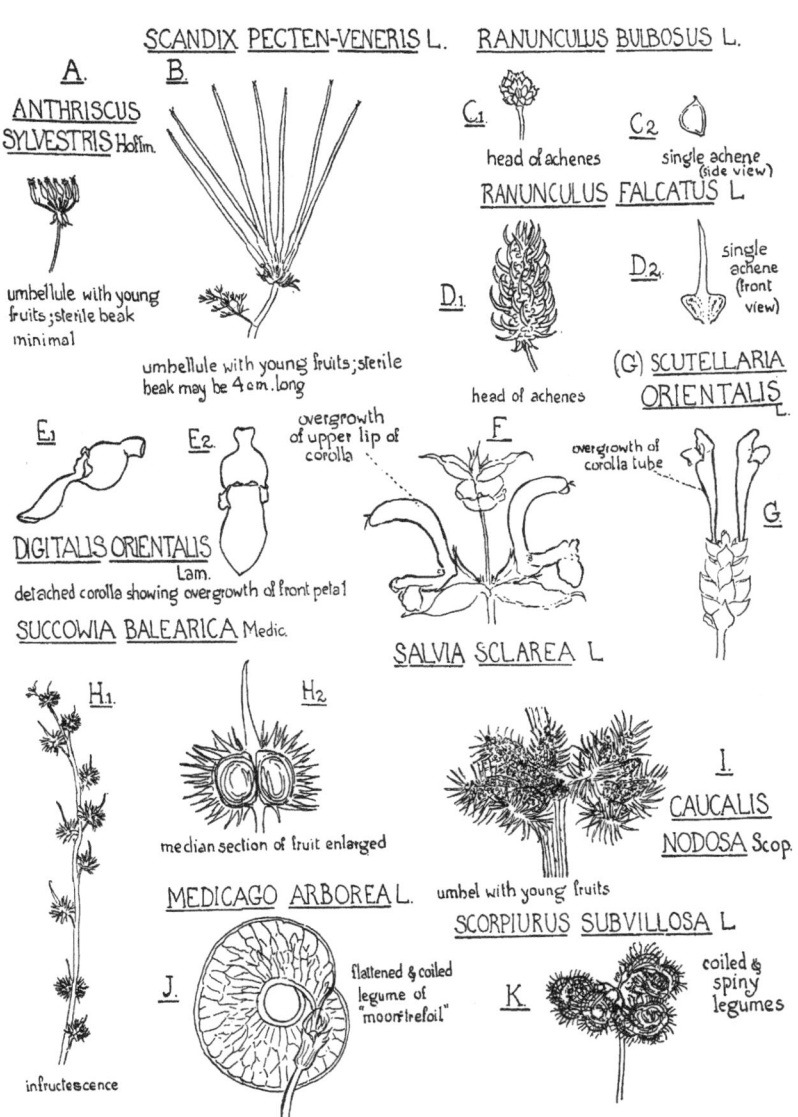

SCANDIX PECTEN-VENERIS L. RANUNCULUS BULBOSUS L.

A.

ANTHRISCUS
SYLVESTRIS Hoffm.

B.

umbellule with young
fruits; sterile beak
minimal

umbellule with young fruits; sterile
beak may be 4 cm. long

C1.

head of achenes

C2

single achene
(side view)

RANUNCULUS FALCATUS L

D1.

head of achenes

D2.

single
achene
(front
view)

E1.

E2.

overgrowth
of upper lip of
corolla

F

overgrowth of
corolla tube

(G) SCUTELLARIA
ORIENTALIS L.

G.

DIGITALIS ORIENTALIS
Lam.
detached corolla showing overgrowth of front petal

SUCCOWIA BALEARICA Medic.

SALVIA SCLAREA L

H1.

H2.

median section of fruit enlarged

MEDICAGO ARBOREA L.

J.

flattened & coiled
legume of
"moon trefoil"

I.

CAUCALIS
NODOSA Scop.

umbel with young fruits

SCORPIURUS SUBVILLOSA L

K

coiled &
spiny
legumes

infructescence

Fig. 35. Differential growth in flower parts. All drawings (×½),
except C2, D2, H2, I, enlarged; J (× 1¾).

169

Fig. 35, C. Examples of such distinctions might be multiplied indefinitely, but we will merely cite the exaggeration of the front petal in *Digitalis orientalis* L. (Fig. 35, E) as compared with that of *D. purpurea* L., and the proportions of the corolla in certain Labiates; in *Scutellaria orientalis* L. (Fig. 35, G) the upper lip is less than half the length of the cylindrical tube, while in *Salvia Sclarea* L. (Fig. 35, F) the upper lip is more than twice the length of the tube.

Huxley's definition of relative growth, already quoted, includes "growth localized at certain circumscribed spots". Of this we may find innumerable cases amongst those gynaecea which become spiny at maturity. As fruits thus distinguished from others of their respective families, we may mention *Succowia Balearica* Medic. (Cruciferae, Fig. 35, H) and *Caucalis nodosa* Scop. (Umbelliferae, Fig. 35, I). The ultimate shapes of fruits may also be much modified by another form of differential growth, in which the region towards one margin of an ovary grows faster than the opposite margin, thus producing coiling; examples may be found among such Leguminous plants as *Medicago arborea* L. (Fig. 35, J) and *Scorpiurus subvillosa* L. (Fig. 35, K).

Occasionally we may glean some hint as to the relations of differential growth; it sometimes appears, for instance, that inhibition in one direction finds its compensation in increased development in another. Inflorescences and flowers offer many examples which lend themselves to this interpretation. It is possible, for instance, that the tendency to reduction or absence of perianth members and bracteoles in the Aroids is directly connected with the relatively large size of the spathe.[1] There seems, also, to be a correlation between elaborate branching of an inflorescence, and reduction in the parts of the individual flower; the grasses, with their much-branched panicles and minimal floral members, offer striking examples.

Turning to internal anatomy, we find that differential growth may play its part in the specialisation of tissues. In the development of the tobacco leaf (*Nicotiana Tabacum* L.), a clear case can be made out for the origin of a characteristic structure through such relative growth. It has been shown that the cell-enlarge-

[1] Engler, A. (1897), p. 378.

ment of the lower epidermis continues longer than that of the internal tissues. The strain thus exerted by the surface layer pulls the mesophyll cells apart, inducing the 'spongy' arrangement, while, correspondingly, the pull of the spongy tissue upon the epidermis, gives the cells of this layer their sinuous outline.[1]

The origin of the sponginess of the mesophyll is an example of a special *internal* structure originating from the physical interaction of tissues which are in contact, so that the cells are subjected to what have been called 'social' influences.[2] In the same way, *external* form may be influenced by the interaction of members which impinge upon one another; the crowding, which is liable to occur in embryonic stages, may, for instance, lead to a failure in disjunction of the young parts. The terms for different kinds of union or non-separation are ill-defined. It is a pity that Robert Brown's hint that the word 'connation' might be the most suitable to indicate what has sometimes been called 'congenital fusion'[3] (failure to achieve freedom in development) has not been adopted; it is this kind of union which is in question here. When two phyllomes occur at one node, they are not infrequently united symmetrically at the base. Such a union has no apparent effect upon succeeding leaves; but a *one-sided* union of two leaves at a node may have an inhibiting influence on the next leaf member, which stands above the line of junction. This is illustrated in the abnormal shoot of sycamore (*Acer Pseudo-platanus* L.) drawn in Fig. 36, A 1–A 3, p. 172. The leaves of one nodal pair were united by one margin of each petiole. The succeeding pair of leaves, alternating with the fused pair, showed no fusion, but the one, which was on the same orthostichy as the junction-line of the two petioles belonging to the node below, was much smaller than its fellow. A more extreme case of a similar effect, but in a normal structure, may perhaps be recognised in the flower of *Plantago*. Here, if we accept the view sponsored by Eichler,[4] the back petal is equivalent to two in a state of union, and, correspondingly, the back stamen, which should occur between the two, is absent.

[1] Avery, G. S. (1933), p. 590.   [2] Macleod, J. (1919, 1926), p. 73.
[3] Brown, R. (1840), pp. 562–3.
[4] Eichler, A. W. (1875, 1878), pt. 1, p. 225.

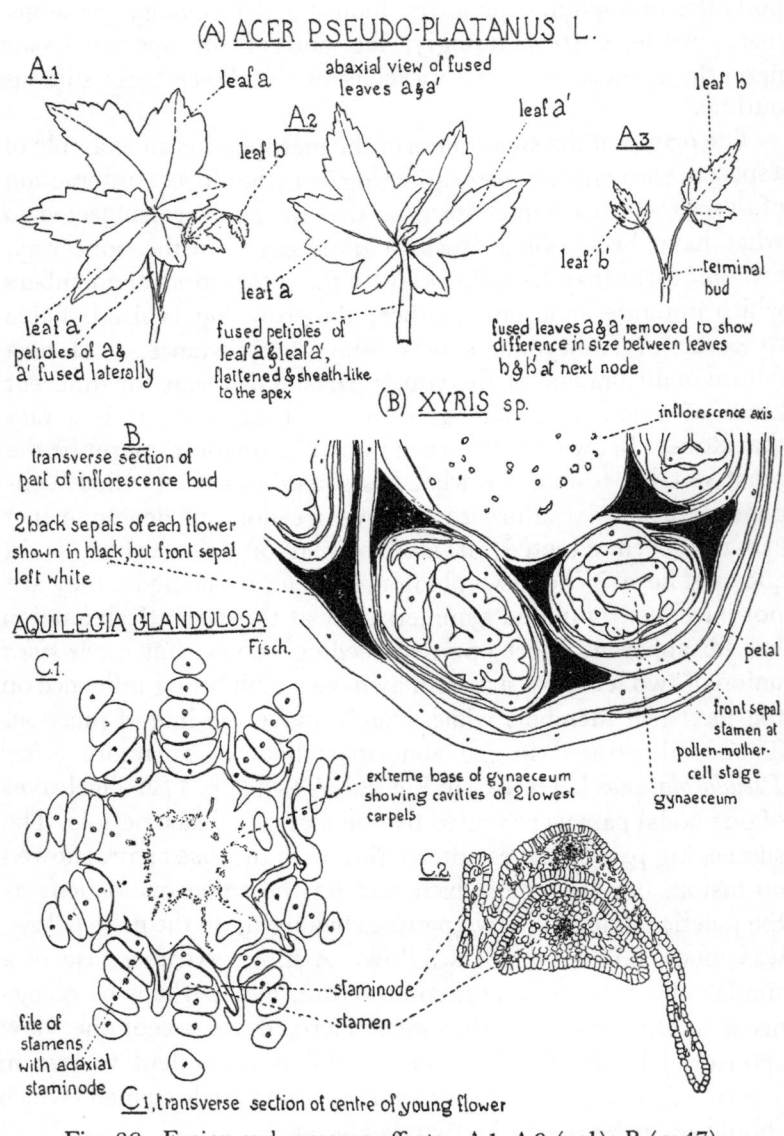

(A) ACER PSEUDO-PLATANUS L.

A.1

leaf a

abaxial view of fused
leaves a & a'

A.2

leaf b

leaf a'

leaf a

A.3

leaf b

leaf a'
petioles of a &
a' fused laterally

fused petioles of
leaf a & leaf a',
flattened & sheath-like
to the apex

leaf b'

terminal
bud

fused leaves a & a' removed to show
difference in size between leaves
b & b' at next node

(B) XYRIS sp.

inflorescence axis

B.
transverse section of
part of inflorescence bud

2 back sepals of each flower
shown in black, but front sepal
left white

petal

front sepal
stamen at
pollen-mother-
cell stage
gynaeceum

AQUILEGIA GLANDULOSA
Fisch.

C.1.

extreme base of gynaeceum
showing cavities of 2 lowest
carpels

C.2

staminode

stamen

file of
stamens
with adaxial
staminode

C.1, transverse section of centre of young flower

Fig. 36. Fusion and pressure effects. A 1–A 3 ($\times\frac{1}{2}$); B ($\times 47$),
lateral sepals black; C 1 ($\times 23$), C 2 ($\times 77$ *circa*).

172

That the shape of various members of the plant body is modified by the members adjacent to them, has long been acknowledged,[1] but we have only circumstantial evidence as to exactly what factor it is that induces the result. Judging merely from inspection, it looks as if limitation of the space into which to expand, and the actual pressure which the developing parts exert upon one another, must be the efficient cause. This impression is confirmed by transverse sections of flower-buds, cut with the microtome, so that the relative position of the parts is preserved. Fig. 36, B, p. 172, shows a section across part of the inflorescence-bud of a species of *Xyris*. Each flower has three sepals—a wider one in front, and two laterals; for clearness the lateral sepals are indicated in black. It will be seen that their triangular sectional form is accurately moulded to the available crannies between the close-packed flowers. The result is that these two sepals, considered together, bear in section a certain resemblance to the bifid palea of the grasses—a member whose form is influenced in the same way by space conditions. In the Gramineae, indeed, pressure in development seems to be a primary factor in relation to form,[2] and equally cogent examples can be found among the dicotyledons. The effects of crowding in the bud stage upon the form of the flower, was long ago studied by Godron for the Cruciferae and Fumarioideae.[3] As an instance from another dicotyledonous family, we may take the staminodes of *Aquilegia glandulosa* Fisch. (Fig. 36, C), which, as seen in a section of a young flower, seem each to be squeezed out laterally between the row of stamens on the same orthostichy, and the gynaeceum. There are indications also that, in *Caucalis nodosa* Scop. (Fig. 35, I, p. 169), the distribution of spines may depend upon space conditions, for these spines are developed chiefly on the outer face of the outer gynaeceum of each partial umbel. Another more important effect of crowding has been noted by Philipson in his study of the capitulum-ontogeny of *Bellis perennis* L. He makes the interesting suggestion that the ray florets owe their incomplete character to lack of space during their early growth.[4]

[1] Goethe, J. W. von (1790): § 36 (cf. p. 44 of present book); Candolle, A. P. de (1813), p. 101.

[2] Arber, A. (1934), pp. 176–80; see also pp. 112–14, 117, 124–7, 133, 206, 408.

[3] Godron, D. A. (1864a, b); see also Arber, A. (1931a), pp. 199–200.

[4] Philipson, W. R. (1946), p. 266.

X. THE MECHANISM OF PLANT MORPHOLOGY

Perhaps the most hopeful field, in which to analyse the dependence of shape upon pressure, is that of ovule development, which has been the subject of a study by Joshi.[1] It is well known that ovules at maturity may be either straight (orthotropous, or, to use a better term, atropous), or bent (anatropous). Ovules are, in general, typically atropous in their earliest stages, but this straightness is, as a rule, only maintained permanently in carpels or gynaecea which are either one- or few-ovuled. As a typical example we may cite the gynaeceum of *Atraphaxis buxifolia* Jaub. et Spach (Polygonaceae), which, with its single erect ovule, is shown in longitudinal section in Fig. 37, A. The solitary state of the ovule, and its basal attachment, save it from any distorting pressure in youth, and it remains permanently atropous. When, on the other hand, we examine those ovules which are bent at maturity, we often find that the gynaeceum contains a large number, which, in the early stage (see Fig. 37, B–D), are packed together with so extreme a closeness as to suggest that their curvature is indeed forced upon them from the outside in the course of development. This view is confirmed by the fact that, in a species in which the ovules, normally, are anatropous at maturity, they may, in exceptional cases, remain atropous, if their formation occurs under conditions of abnormal freedom. For instance, in a mutant of *Primula sinensis* Sabine, it has been recorded[2] that the stamens were leafy and bore ovules, which were more or less anatropous if the stamen was folded and the ovules were thus situated in a cavity, but were, on the other hand, generally atropous when they were produced on an open surface.

In the development of anatropous ovules we have a typical example of departure from an originally radial symmetry. In considering the part played in morphology by the concept of symmetry, it must be recalled that the accepted meaning of this term is now narrower than formerly.[3] At the beginning of the nineteenth century Corréa da Serra gave a very broad definition. He explained, "J'entends par symétrie l'arrangement particulier des parties qui résulte de leur situation respective et de leurs formes."[4] De Candolle, writing a little later, understood symmetry to mean the "système général de l'or-

[1] Joshi, A. C. (1935), pp. 288–9. [2] Brieger, F. G. (1935).
[3] Lovejoy, A. O. (1948), p. 146. [4] Corréa da Serra, J. F. (1805), p. 377.

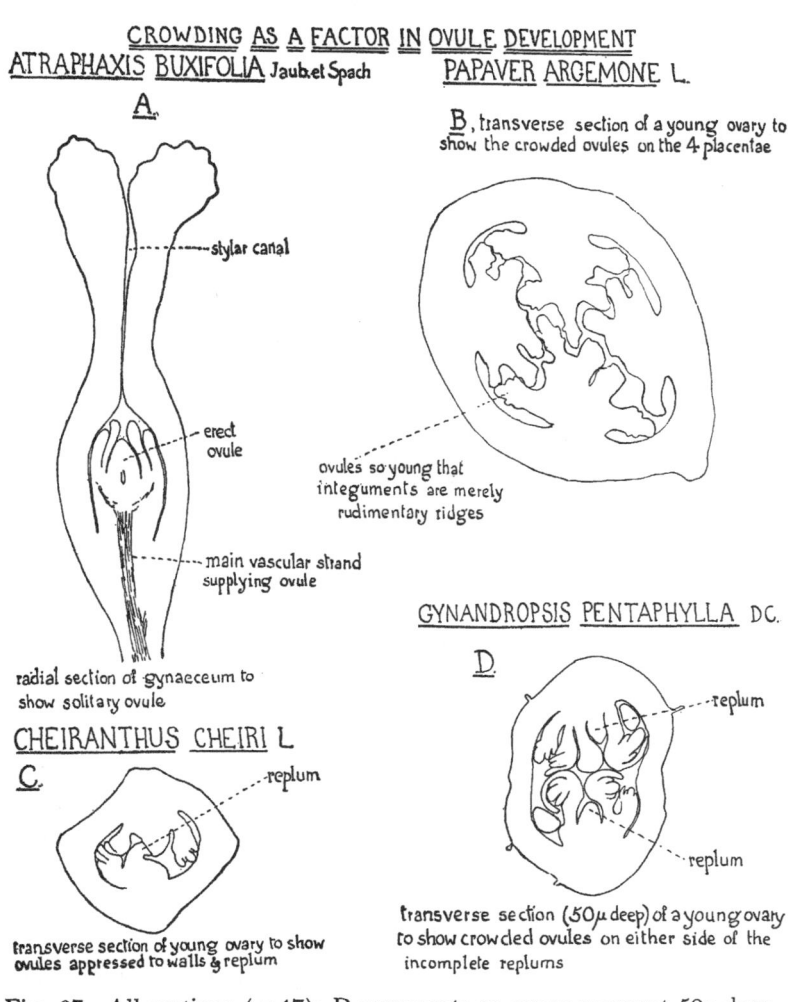

CROWDING AS A FACTOR IN OVULE DEVELOPMENT

ATRAPHAXIS BUXIFOLIA Jaub.et Spach

PAPAVER ARGEMONE L.

A.

B, transverse section of a young ovary to show the crowded ovules on the 4 placentae

stylar canal

erect ovule

ovules so young that integuments are merely rudimentary ridges

main vascular strand supplying ovule

radial section of gynaeceum to show solitary ovule

GYNANDROPSIS PENTAPHYLLA DC.

D.

replum

CHEIRANTHUS CHEIRI L

C.

replum

replum

transverse section of young ovary to show ovules appressed to walls & replum

transverse section (50μ deep) of a young ovary to show crowded ovules on either side of the incomplete replums

Fig. 37. All sections (× 47); D represents an ovary segment 50μ deep.

175

ganisation",[1] but he also defined it more specifically as "cette régularité non géométrique des corps organisés".[2] In considering the problems discussed in the present book, we have often had occasion to refer to repetitive growth, and at this point the idea inevitably recurs, for—just as rhythm is the outcome of repetition in time—symmetry is the outcome of spatial repetition.[3] Radial symmetry characterises the stationary plant, in conse-

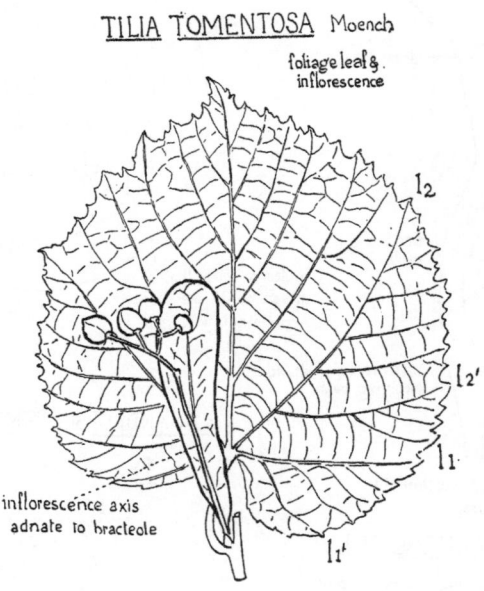

TILIA TOMENTOSA Moench

foliage leaf &
inflorescence

l2

l2'

l1

inflorescence axis
adnate to bracteole

l1'

Fig. 38. *Tilia tomentosa* Moench. Foliage-leaf seen from below with its associated inflorescence (× ½); *l* 1 and *l* 2, main lateral veins; *l* 1′ and *l* 2′, secondary lateral veins.

quence of its mode of growth, just as bilateral symmetry, with a directional sense, characterises the moving animal. De Candolle's personal contribution to the theory of symmetry was his idea that "tous les êtres organisés sont réguliers dans leur nature intime";[4] he claims to have been one of the first to state the principle that organised beings are symmetrical or regular

---

[1] Candolle, A. P. de (1813), p. 92.
[2] Candolle, A. P. de (1827), vol. II, p. 238.
[3] Woodger, J. H. (1929), p. 306.   [4] Candolle, A. P. de (1813), p. 102.

when considered in their type.[1] What he means by this is illustrated by his statement that the zygomorphic Scrophulariaceous flower is a modification of the Solanaceous type, "parce qu'une Personée régularisée par la pensée ne diffère pas d'une Solanée".[2] If this principle of original basic symmetry be accepted, it remains to account for the passage from the hypothetical regular prototype to the non-radial plant forms which actually exist. De Candolle's solution of the problem is to explain departures from radial symmetry as due to specific causes—"accidens plus ou moins constans"[3] which lead to such changes as abortion, degeneration, fusion, or multiplication of parts. Among these 'accidens', de Candolle allots the chief role to relative position within the shoot system; for though the repetition of shoot generations involves a certain general symmetry, it is also true that the fact that these generations remain permanently attached to one another, means that growth is often cramped on the face by which each is connected with its parent, and that some asymmetry thus results. This is illustrated over and over again in the venation of leaves. In such a lamina as that of the lime (*Tilia*) drawn in Fig. 38, p. 176, the two main laterals arising at the base of the blade on the right-hand side (l_1 and l_2), themselves bear a series of strong laterals of the second order (l_1', l_2', etc.), on the face remote from the midrib—that is to say, on the side on which the leaf can develop freely, while, on the other side, the venation is comparatively feeble. The production of lobes or pinnules from pinnae also tends to be restricted to the side towards the base of the whole leaf; this is seen, for instance, in *Chelidonium majus* L. (Fig. 18, E, p. 115) and in *Angelica ampla* Nelson (Fig. 18, B).

One of the ways in which relative position in the branch system may influence form, is through increasing or diminishing the inflow of sap which is needed for growth. Changes which are probably related to an increased flow of nutriment are witnessed when woody plants are hedged or coppiced, so that fresh shoots arising from the base are in a privileged position in regard to the roots. For example, when the hazel (*Corylus Avellana* L.) is cut

---

[1] Candolle, A. P. de (1827), vol. II, p. 239.
[2] Candolle, A. P. de (1813), p. 144.
[3] Candolle, A. P. de (1827), vol. I, p. 519.

X. THE MECHANISM OF PLANT MORPHOLOGY

back, the flowers of the female catkins on the new and vigorous shoots may be produced in groups of three; that is to say the median flower, which occurs in the related *Betula,* but is suppressed normally in *Corylus,* may be developed. The bract, also, subtending each group of flowers, may be tripartite instead of simple.[1] The results of such crude and unintentional experiments are suggestive in connexion with certain features of other life-histories, which have not been disturbed artificially—for instance, the rhythm of phyllome production at a shoot apex. During the period in which vigorous shoot extension monopolises much of the available food, the new primordia developed at the tip of a leafy shoot are—owing to their position above the extending region—poorly placed for supplies; the result is that they are able to form only bud-scales. On the other hand, when extension ceases, more complex phyllomes (foliage-leaves) can again be produced.[2] In reproductive shoots, certain changes can, with some probability, be assigned to differences in food supply due to relative position. The basal flower of a raceme, for instance, since it has the first call upon the flow of sap, sometimes shows a supernormal luxuriance. The peloric flowers[3] of *Antirrhinum majus* L. and of *Linaria vulgaris* Mill., drawn in Fig. 44, A2 and C, p. 189, were the basal flowers of racemes, and each had six petals instead of five. In Fig. 39, A1–A3, p. 179, again, the lowest of 25 flowers in a raceme of *Aconitum Napellus* L. (monkshood) is seen to be accompanied by two supernumerary flowers, arising in the axils of its two bracteoles, which normally are sterile. In such a development, which is comparable with the extra flower production in the coppiced hazel already cited, there is obviously nothing monstrous—what happens is merely a replacement of a simple scheme of branching by one that is more complex; the hypertrophy of basal flowers may, on the other hand, disturb the balance of development, and express itself in non-functional structures, intermediate between flowers and inflorescences. In the first-formed members of the raceme of water-cress (*Nasturtium officinale* R.Br.), for example, a curious anomaly occasionally occurs (Fig. 39, B, p. 179): one to four

[1] Weiss, F. E. (1932).
[2] Underwood, D. & Scott, L. I., in Priestley, J. H. and Scott, L. I. (1935), p. 219.
[3] See pp. 185 *et seq.*

178

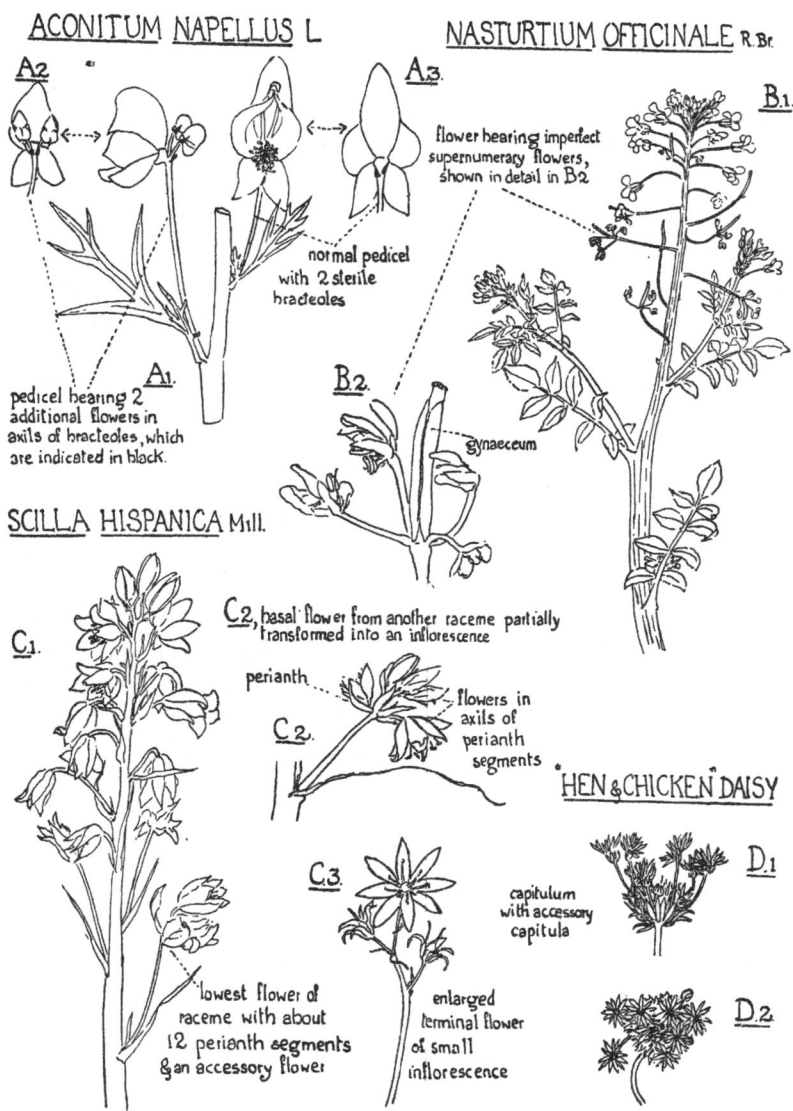

ACONITUM NAPELLUS L

A2

A3

flower bearing imperfect
supernumerary flowers,
shown in detail in B2

normal pedicel
with 2 sterile
bracteoles

A1.

pedicel bearing 2
additional flowers in
axils of bracteoles, which
are indicated in black.

NASTURTIUM OFFICINALE R.Br.

B.1.

B2.

gynaeceum

SCILLA HISPANICA Mill.

C.1.

C2, basal flower from another raceme partially
transformed into an inflorescence

perianth

flowers in
axils of
perianth
segments

C 2.

"HEN & CHICKEN" DAISY

C.3.

capitulum
with accessory
capitula

D.1.

lowest flower of
raceme with about
12 perianth segments
& an accessory flower

enlarged
terminal flower
of small
inflorescence

D.2

Fig. 39. Inflorescences replacing flowers. All drawings (×½)
except B2 (enlarged).

179

imperfect accessory flowers arise from the parent flower, each deriving its vascular supply from a petal strand.[1] Transitional forms between flowers and inflorescences may also be found in *Scilla hispanica* Mill. growing under garden conditions. Fig. 39, C 1 shows an inflorescence in which the lowest flower had about 12 perianth segments, in the axil of one of which there was another flower. In Fig. 39, C 2, the lowest 'flower' from another similar raceme is drawn. Here the bract, bracteole, and pedicel suggested a normal flower, but only two or three perianth segments were present, and they were succeeded by a little inflorescence of seven flowers. It is sometimes assumed that garden life 'causes' such deviations as these, but it is possible that, in certain cases, cultivation does no more than reveal something which is present, though inconspicuously, in the wild state. This seems to be true of the 'hen-and-chicken' daisy, an abnormality which bears some affinity to those just cited, since extra inflorescences are produced from the outer (morphologically lower) region of the capitulum. An example, which was found among normal daisies flowering in a pasture field in Dorset some years ago, looked different from its neighbours only in a slight prominence of the involucral bracts. This appearance was found to be due to minute, embryonic, secondary capitula, occupying the axils of the involucral bracts, and forcing them a little outwards. After the plant had been dug up and transferred to a garden, further flower heads showing this character were produced in the same season; the stalks of the secondary capitula still did not elongate, so that the abnormality remained inconspicuous. During the next spring, however, in the first two flowerheads, which were specially vigorous, the stalks of the secondary capitula elongated, producing the effect shown in Fig. 39, D 1, D 2. This abnormality was thus in no sense *due* to cultivation, but garden conditions merely gave it the opportunity of displaying itself more obviously than in the field.

It is not only basal flowers which have a tendency to overgrowth; this tendency may also be shown by apical flowers, which may occupy an equally privileged position as regards the main stream of sap. Fig. 39, C 3, p. 179, represents a small three-flowered inflorescence of *Scilla hispanica* Mill., in which

[1] Arber, A. (1931 a), pp. 187–95.

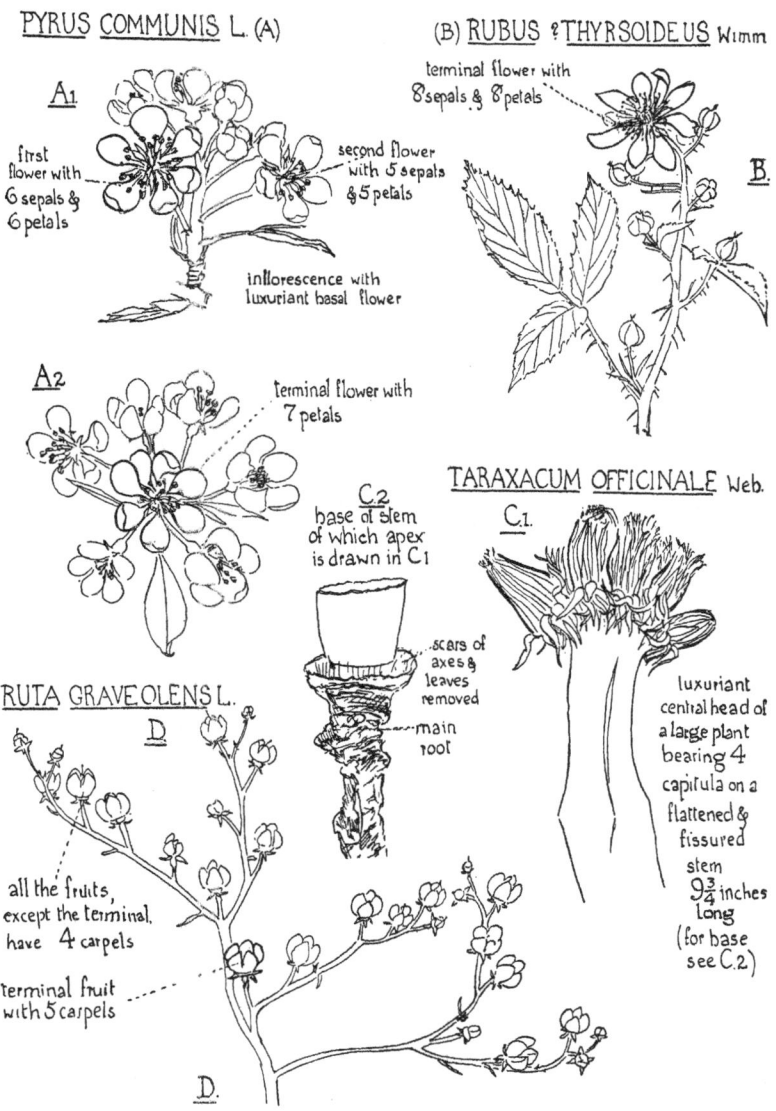

PYRUS COMMUNIS L. (A)

A1

first flower with 6 sepals & 6 petals

second flower with 5 sepals & 5 petals

inflorescence with luxuriant basal flower

A2

terminal flower with 7 petals

(B) RUBUS ?THYRSOIDEUS Wimm

terminal flower with 8 sepals & 8 petals

B.

TARAXACUM OFFICINALE Web.

C.1.

C.2
base of stem of which apex is drawn in C1

scars of axes & leaves removed

main root

luxuriant central head of a large plant bearing 4 capitula on a flattened & fissured stem $9\frac{3}{4}$ inches long (for base see C.2)

RUTA GRAVEOLENS L.

D

all the fruits, except the terminal, have 4 carpels

terminal fruit with 5 carpels

D.

Fig. 40. Hypertrophy of basal or terminal flowers or inflorescences.
All drawings ($\times\frac{1}{2}$).

181

there is a terminal flower with eight perianth members and eight stamens. Again, in the two inflorescences of the pear (*Pyrus communis* L.) drawn in Fig. 40, A 1, A 2, p. 181, one has a basal flower with six sepals and six petals, while the other has a terminal flower with six sepals and seven petals. Luxuriance, associated with central placing, is sometimes seen in the dandelion (*Taraxacum officinale* Web.); a plant in which this was observed had 86 leaves and 28 inflorescences, surrounding the fasciated four-headed shoot drawn in Fig. 40, C 1, p. 181. An overgrown umbel of the carrot (*Daucus Carota* L.), borne at the end of the main axis, is sketched in Fig. 28, p. 150. The relatively large scale of the terminal inflorescence in the teasle used for dressing cloth (*Dipsacus fullonum* L.), is so constant a character as to have been of industrial importance. The plant may be as much as seven feet high, and may bear 40 to 100 heads; of these the one terminating the main axis is so much larger and finer than the rest that it used to be distinguished as a 'king', while the others were known as 'middlings' and 'scrubs'.[1] Not only in inflorescences, but also in individual flowers, an increase in the number of parts may be quite a regularised event at the apex of the reproductive shoot. It is a character, for instance, of the form of blackberry known as *Rubus thyrsoideus* Wimm.;[1] the inflorescence with a terminal eight-petalled flower, drawn in Fig. 40, B, p. 181, probably belongs to this species. Another familiar case is that of rue (*Ruta graveolens* L.), of which a fruiting shoot is sketched in Fig. 40, D; the central flower of the partial inflorescence has its parts in fives, while those of the other flowers are in fours. In the inflorescence of *Euphorbia helioscopia* L., the terminal cyathium (flower-like partial inflorescence) has five nectariferous bracts, while the laterals have four only; here the character of the terminal *cyathium* corresponds with that of the terminal *flower* in *Ruta*[5] (Fig. 40, D).

An example, in which a variant of a slightly different kind is associated with the terminal position, is found in the flowering shoot of the carrot (*Daucus Carota* L.). It is not easy to determine the class of inflorescence to which any given form of umbel

[1] Hennell, T. (1936), p. 192.
[2] Velenovský, J. (1905–13), vol. III, 1910, p. 858.
[3] Delpino, F. (1869), p. 113.

THE TERMINAL UMBELLULE OF <u>DAUCUS</u> <u>CAROTA</u> L.

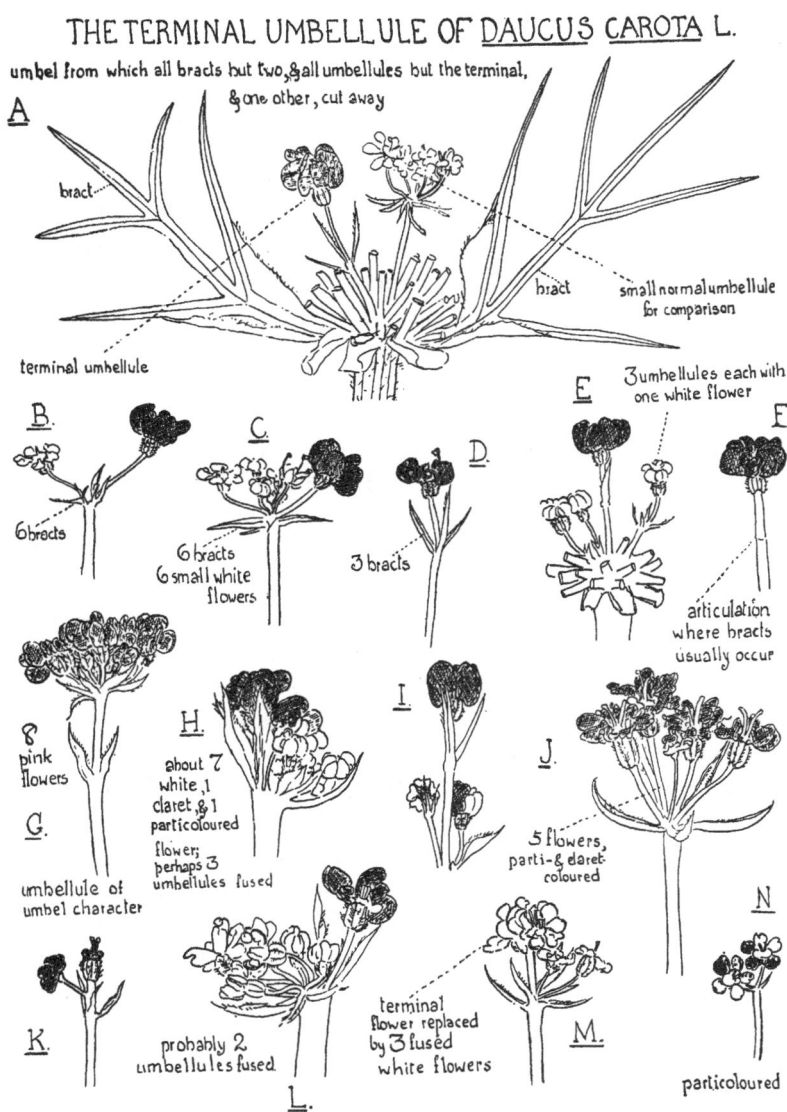

umbel from which all bracts but two, & all umbellules but the terminal, & one other, cut away

A

bract

bract

small normal umbellule for comparison

terminal umbellule

B.

6 bracts

C.

6 bracts 6 small white flowers

D.

3 bracts

E

3 umbellules each with one white flower

F.

articulation where bracts usually occur

G.

8 pink flowers

umbellule of umbel character

H.

about 7 white, 1 claret, & 1 particoloured flower; perhaps 3 umbellules fused

I.

J.

5 flowers, parti- & claret-coloured

N.

K.

L.

probably 2 umbellules fused

terminal flower replaced by 3 fused white flowers

M.

particoloured

Fig. 41. From plants collected near Cambridge. Drawings enlarged.

should be referred, but there seems to be evidence that those of the Umbelliferae[1] are really cymose. The central umbellule in the carrot, which would thus be terminal, shows a range of peculiarity. In its commonest form it is represented by a single flower, which, however, may be interpreted as a reduced umbellule, since there are usually one or more bracts part way up the stalk, recalling those at the base of the normal umbellule. This single flower is markedly larger than the normal flowers, and is often claret-coloured, red, or pink, so that it contrasts strikingly with the white flowers of the surrounding umbellules. A selection of the forms taken by the central umbellule is illustrated in Fig. 41, p. 183; it may have more than one flower, and these may differ in colour, or may even be individually parti-coloured. Rarely the number of coloured flowers is high, thus approximating to the condition found in another species, *D. Gingidium* L. There were eight pink flowers in the specimen drawn in Fig. 41, G, and the distribution of bracts suggests that, in this case, the central umbellule tended to transformation into an umbel. The flower of the central umbellule, when only one is present, agrees with those other examples of terminal flowers which we have noticed, in its exaggerated size. Moreover, its coloration may also be taken as symptomatic of excess of nourishment, for accumulation of synthetic products is known to favour the development of anthocyanin.[2] As regards petal size and pigmentation, the carrot thus illustrates overgrowth of a terminal flower, but it also—in the reduction of the terminal umbellule to the one-flowered condition—exemplifies the opposite tendency: it shows, that is, the reduction of a terminal structure, presumably owing to the demands of the laterals below it.[3] In plants in general, this reduction of the apical part of the main axis operates especially when the laterals below are paired or whorled, for this arrangement necessarily involves the all-round tapping of the main food supply immediately before the terminal member has the opportunity of drawing upon it. An example comparable with that described for the carrot, but represented in a different order of branching, is revealed by the lay-out of the flowering plant of

[1] Uittien, H. (1928a), pp. 410–11.
[2] Wheldale, M. (1916), pp. 84–5.
[3] This subject has been discussed in general in chapter VII.

Queen-Anne's-lace (cow-parsley, *Anthriscus sylvestris* Hoffm.). The main axis is generally found to conclude with a small umbel, reduced to a few umbellules, or, more commonly, to a single one (Fig. 42, p. 186). Immediately below this reduced terminal structure, there is usually a whorl of three lateral reproductive shoots, or else a pair. Unlike the terminal umbel, these shoots are fully developed; they each end in an umbel, and also bear lateral umbels. One can easily see that their demands, as they grow, must militate against an adequate flow of sap passing to the terminal umbel. This tendency to terminal reduction, which appears in *Daucus*, *Anthriscus*, and *Euphorbia*, in a complex of inflorescences, may be expressed also within a single inflorescence. In the five-faced-counsellor[1] (*Adoxa moschatellina* L.), the 'capitulum-like cyme'[2] ends in a flower which has to compete with four flowers, in two decussating pairs, immediately below it. It is four-partite, whereas the lateral flowers have their parts in fives.

So far we have touched only upon changes in size, vigour, and number of parts connected with the position of the flower or inflorescence, but changes of a more strictly morphological kind may also be associated with situation. Zygomorphic flowers commonly show their freest development in the direction away from the inflorescence axis; the part of the corolla backing on the axis is relatively reduced, while the front part is expanded to form the so-called labellum or lip, which is often distinguished by special colouring or markings. There are, however, deviations from this kind of zygomorphy, which one hesitates to call abnormalities, since they show a markedly law-abiding structure, and their location is far from haphazard. These deviations, which have been known as *peloria*, ever since detailed attention was first called to them by a pupil of Linnaeus,[3] all fall into one category, in being radially symmetrical instead of zygomorphic, and in attaining their symmetry all in the same way, through building their whole corolla by repetition of the part that in the zygomorphic flower is the front lip. The interest of the peloric form, from our present standpoint, is that its occurrence in place

[1] Cited as a local name in McDonald, J. E. (1904).
[2] Sprague, T. A. (1927), p. 472.
[3] Rudberg, D. (1744), in Linnaeus, C. (1749), *De Peloria*, pp. 280–98.

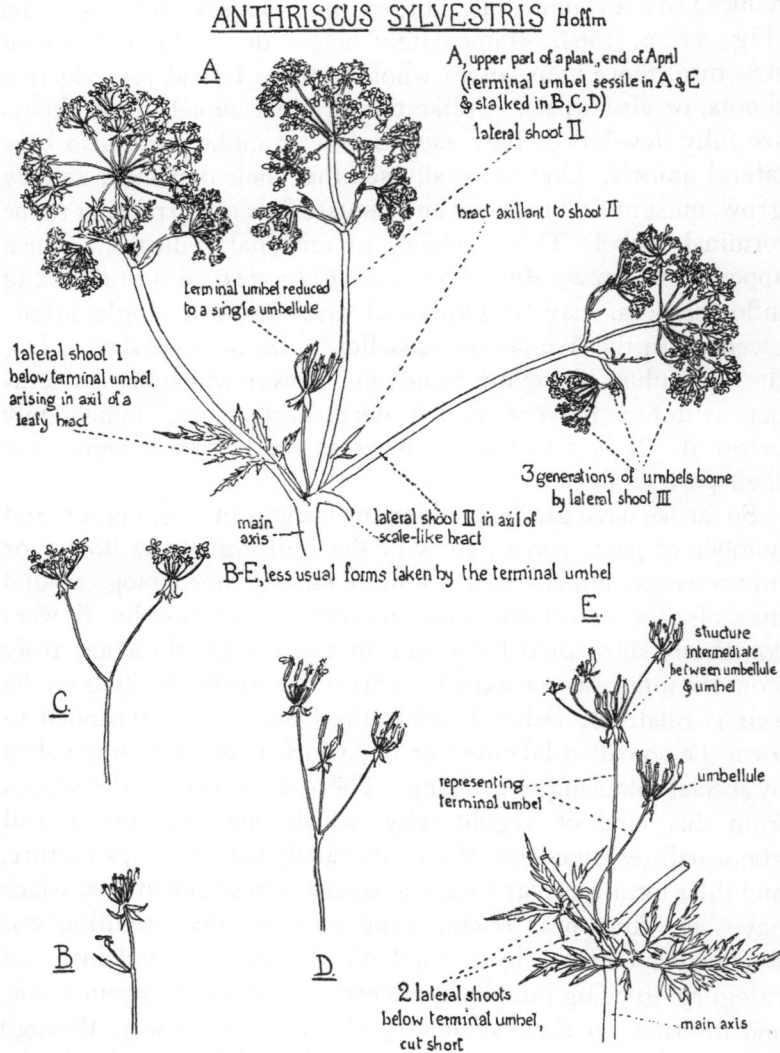

ANTHRISCUS SYLVESTRIS Hoffm

A, upper part of a plant, end of April
(terminal umbel sessile in A & E
& stalked in B, C, D)

lateral shoot II

bract axillant to shoot II

terminal umbel reduced
to a single umbellule

lateral shoot I
below terminal umbel,
arising in axil of a
leafy bract

3 generations of umbels borne
by lateral shoot III

main
axis

lateral shoot III in axil of
scale-like bract

B-E, less usual forms taken by the terminal umbel

C.

E.

structure
intermediate
between umbellule
& umbel

umbellule

representing
terminal umbel

B

D

2 lateral shoots
below terminal umbel,
cut short

main axis

Fig. 42. Termination of main shoot. All drawings (×½).

186

of the zygomorphic form is frequently related to situation in the shoot system; when this is so, we may regard zygomorphy and actinomorphy as functions of position. A terminal situation in the inflorescence, with the consequent freedom from pressure against the axis—and with the facilities which this apical position affords for direct tapping of the main stream of food material— seems to favour symmetrical peloric forms, while the lateral position is associated with zygomorphy. Some of the best known types of peloria are those not infrequently produced in the foxglove (*Digitalis purpurea* L.). Here the raceme may develop at its tip a cup-shaped flower, in which the petals may be free. This flower is not merely a radially symmetrical version of the ordinary zygomorphic type; it often offers an extreme example of the luxuriance which we have already noticed in certain terminal flowers. In the plant illustrated in Fig. 43, A–D, p. 188, the corolla was white, but, in the terminal flower, each petal showed the spotted character which, in the zygomorphic flowers, is confined to the labellum. Near the top of the inflorescence there are two flowers (Fig. 43, B and C), which are also peloric, though diverging less than the terminal flower from the zygomorphic type. One has five lip-like petals, and five stamens, while the other has three lip-like petals and three stamens. A similar flower with eight lip-like petals, which occurred at the tip of another inflorescence, is seen from above in Fig. 43, E. The developments just below the terminal peloria suggest that the flowers in its proximity have been sacrificed to it. They are represented merely by a crowd of bracts or sepals, some partially petaloid, and by two or three small flowers, so far reduced that they are non-functional. One of these is drawn in Fig. 43, D; it consists of a single lip-petal in tubular form.

Another member of the Scrophulariaceae in which peloric flowers are liable to occur is the snapdragon (*Antirrhinum majus* L.). The plant from which Fig. 44, A3, A4, p. 189, was drawn, had inflorescences which were shorter than usual, and were each terminated by a peloric flower. In this particular case, as in *Digitalis*, the peloric symmetry is thus associated with the terminal position, but this is not always so. Fig. 44, A1, shows a reproductive shoot of another garden variety of *Antirrhinum*, in which the lateral flowers are all peloric, and the same thing is

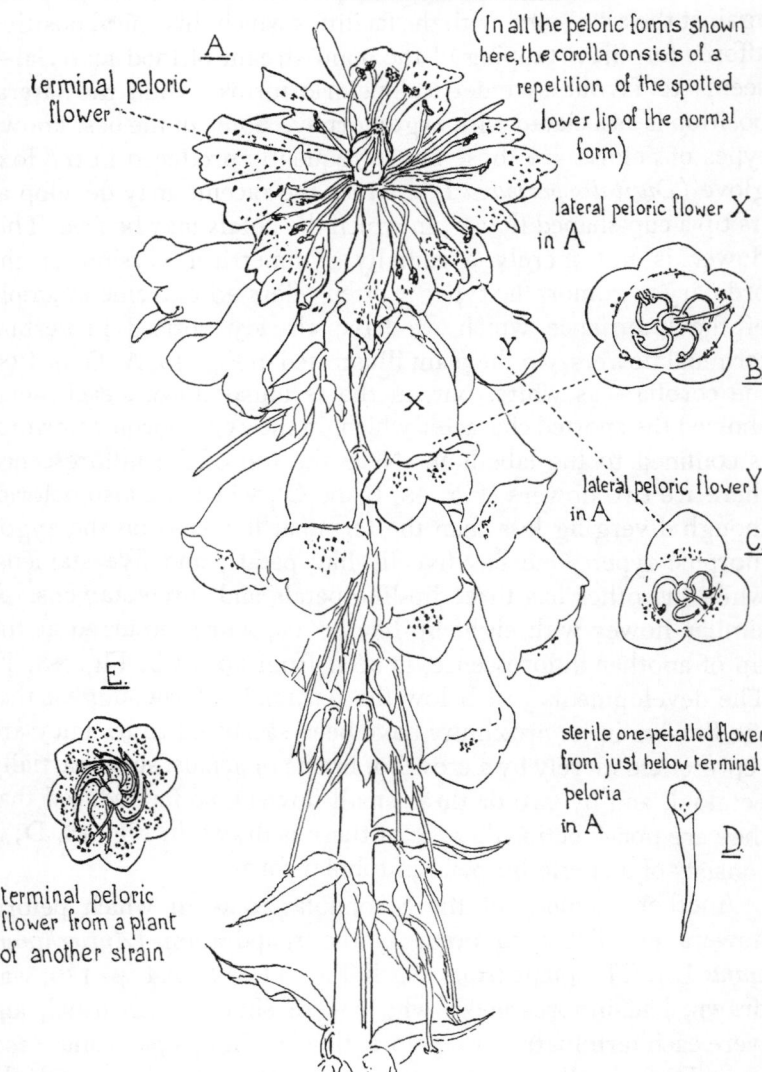

PELORIA OF <u>DIGITALIS</u> <u>PURPUREA</u> L.

A.
terminal peloric flower ·······

(In all the peloric forms shown here, the corolla consists of a repetition of the spotted lower lip of the normal form)

lateral peloric flower X in A

B.

Y

X

lateral peloric flower Y in A

C.

E.

sterile one-petalled flower from just below terminal peloria in A

D.

terminal peloric flower from a plant of another strain

Fig. 43. All drawings (×½).

PELORIA
(A) ANTIRRHINUM MAJUS L.

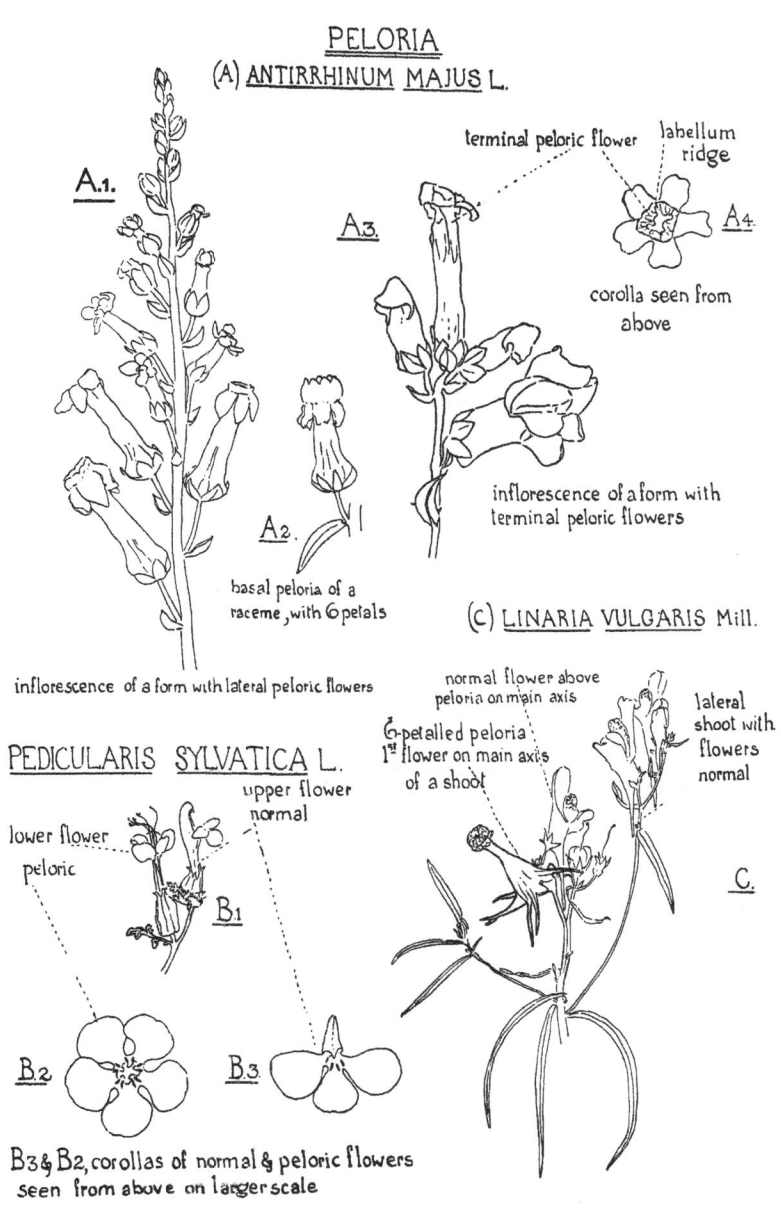

A.1.

A3.

terminal peloric flower   labellum ridge

A4.

corolla seen from above

A2.

basal peloria of a raceme, with 6 petals

inflorescence of a form with terminal peloric flowers

inflorescence of a form with lateral peloric flowers

(C) LINARIA VULGARIS Mill.

PEDICULARIS SYLVATICA L.

lower flower peloric

upper flower normal

B1

normal flower above peloria on main axis

6-petalled peloria 1st flower on main axis of a shoot

lateral shoot with flowers normal

C.

B2   B3

B3 & B2, corollas of normal & peloric flowers seen from above on larger scale

Fig. 44. All drawings (×½) except B2, B3 (slightly enlarged).

189

seen in the engraving accompanying Rudberg's original description of peloria in the related genus *Linaria*.[1] In Fig. 44, again, C and B illustrate the occurrence of an individual non-terminal peloria in *Linaria vulgaris* Mill. and *Pedicularis sylvatica* L. It thus becomes clear that, though there is sometimes a relation between the peloric form and the terminal position, this relation is not obligatory.

In the examples just cited, we have been concerned with peculiarities confined to certain members within the plant body; but the control of shape by physical factors can be regarded also from a more general standpoint. It has long been recognised that there is some connexion between the form of a living creature and its magnitude. Obvious as this may seem, it has often been overlooked, despite the vivid description penned, long ago, by Sachs,[2] of the lethal drawbacks to which a plant of *Marchantia* would be subjected, if it were enlarged to 50 times its linear dimensions, or correspondingly diminished. One special connexion between scale and structure emerges from the physical fact that, when an organ enlarges, its surface area does not keep pace with its increase in volume. Contact with the environment is of such physiological importance, that, as size increases, healthy life may become impossible, unless the form is modified in such a way as to give a relative increase of surface area. Such conditioning of shape by the 'Size-Factor', as Bower[3] calls it, has been demonstrated especially for the primary conducting tracts of the more primitive vascular plants, but it can also be detected elsewhere.

Hitherto we have considered, in the main, only those alterations of form associated with *physical* influences. We saw, however, that in the carrot there was a relation between flower position and a change in *chemistry*, giving rise to anthocyanin production. In flowers in general the differentiation of the parts has been related to chemical factors by the suggestion that the apex of the bud may be visualised as divided into zones respectively carrying hormones for *calyx*, for *petals + stamens*, and for

[1] Rudberg, D. (1744), in Linnaeus, C. (1749), *De Peloria*, pp. 280–98.

[2] Sachs, J. von (1893, vi), pp. 56, etc.

[3] This subject, treated in Bower, F. O. (1923), has been worked out in detail in Bower, F. O. (1930), in which much use is made of Wardlaw, C. W. (1924–8). See also Bower, F. O. (1931), and Ashby, E. (1937).

carpels.[1] Moreover the recurrence of the flower *Gestalt*, in the guise either of a flower or of an inflorescence, indicates that what underlies the *Gestalttypus* would be "parallel sequences of physico-chemical processes leading to similar ends: hormones activating successively produced identical or very closely allied chemical substances at the corresponding stage of each sequence".[2] An example in which such parallel chemical changes become, as it were, visible to the eye, is the passage from peripheral blue to central yellow in simple flowers, such as those of the forget-me-not (*Myosotis*), and, correspondingly, in capitula of the composites, such as those of various asters.[3]

Within the flower itself, it is possible that certain features, such as colour, scent, fragility, and low anatomical differentiation, in which petals are unlike vegetative parts, may be related to some chemical disturbance marking the change-over from the vegetative to the reproductive region. Goethe had perhaps caught a distant glimpse of a truth when he wrote: "It is a very probable notion that the colour and scent of petals are to be attributed to the presence in them of the male fertilising substance."[4] It has been suggested that the essential oils, which give rise to flower scents, may perhaps be regarded as waste products of decomposition, which are likely to be inimical to life,[5] so that, conceivably, the ephemeral character of petals is due to a kind of self-poisoning.

It may be worth remembering that de Candolle[6] held that the flower axis, in the region succeeding the calyx, bore a localised coating, which served as a basis for the corolla and stamens, and which owed its origin to abortion or partial development of these organs. To this growth upon the surface of the male region of the axis he applied the term 'torus', and he used this conception to explain various puzzling features of flower structure. He thought, for instance, that the torus formed the lining of the cup in perigynous flowers, and that elsewhere it provided honey glands and

---

[1] See, for example, Brieger, F. G. (1935), p. 129, and Zimmermann, W. (1935), p. 322; earlier suggestions of such ideas will be found in Sachs, J. (1893, VII), pp. 235–6, and Penzig, O. (1890, 1894), preface to vol. II, 1894, p. v.

[2] Letter from Professor A. G. Tansley, F.R.S., 25 April 1937.

[3] See p. 147. [4] Goethe, J. von (1790), §45.

[5] Hampton, F. A. (1925).

[6] Candolle, A. P. de (1827), vol. I, pp. 483–90.

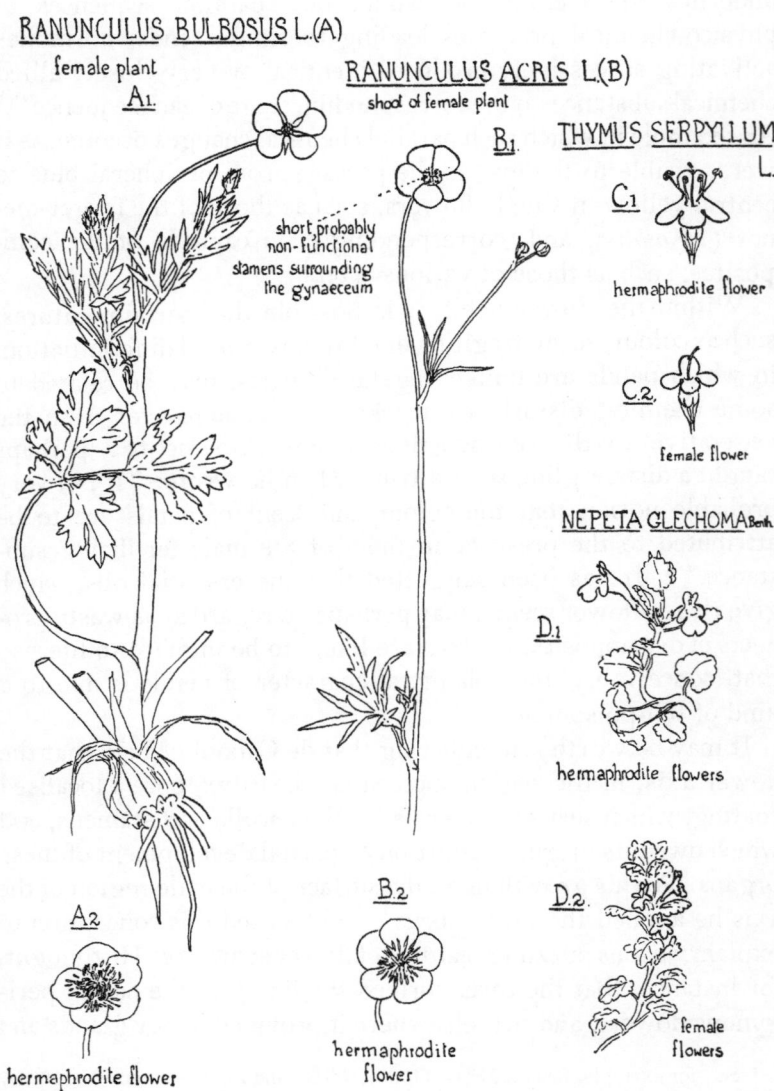

RANUNCULUS BULBOSUS L.(A)
female plant
A₁.

RANUNCULUS ACRIS L.(B)
shoot of female plant

B₁.

THYMUS SERPYLLUM
L.

C₁

hermaphrodite flower

C₂

female flower

short probably
non-functional
stamens surrounding
the gynaeceum

NEPETA GLECHOMA Benth

D₁

hermaphrodite flowers

D₂

female flowers

A₂

hermaphrodite flower

B₂

hermaphrodite flower

Fig. 45. Corolla differences associated with sex differences.
A, B, D (×½); C (enlarged).

192

various expansions of a staminodial character. In *Nymphaea* he considered that the torus adhered to the ovary, and that the stamens were carried up upon it. He describes it in the Passifloraceae as fused with the base of the calyx, which it carpets with a petaloid layer, and as also producing several sets of coloured threads; he believed that the stamens arose from its prolongation over the base of the ovary. For the occurrence on the floral axis of a male surface layer—the torus—we can find an exact parallel among inflorescences. The smooth coating of the apical club of *Arum* has been shown convincingly by Engler[1] to be equivalent to a fused association of abortive staminate flowers, the individuality of which has been wholly lost. This is another instance of the *Gestalt* parallelism between flower and inflorescence discussed in chapter IX. De Candolle's theory of the torus is certainly a stimulus to thought, and, even if its detailed application be not accepted, it has a value in emphasizing the existence of a male *region* in the flower.

If the theory is well-founded, that the passage from sterile to male, and from male to female regions in the flower is accompanied by specific chemical changes, corresponding chemical differences must be held to exist among plants which, as individuals, show differentiation between the sexes, or between fertility and sterility. With these sex differences, certain differences in appearance are often correlated, though such variations are generally a matter merely of size and vigour, and affect the actual form only indirectly. It may possibly be that, in the Compositae, the exaggeration of the corolla in the ray florets, as compared with the disk florets, is a symptom of whatever chemical change underlies the sexual incompleteness of the ray region. Fig. 45, p. 192, shows the distinction between bisexual and female flowers in *Ranunculus*, *Thymus*, and *Nepeta*. In inflorescences, the change to sterility may be associated with a form characterised by abnormal elongation and branching. This is strikingly shown in the 'plume' abnormality of *Plantago major* L., drawn in Fig. 46, C, p. 194; here the close, fertile, plantain spike was replaced by a lax, infertile system, including over 260 slender laterals, which themselves branched. Moreover when, as in *Asparagus* (Fig. 46, B) and in *Eucomis*, a single shoot passes

[1] Engler, A. (1884), p. 299.

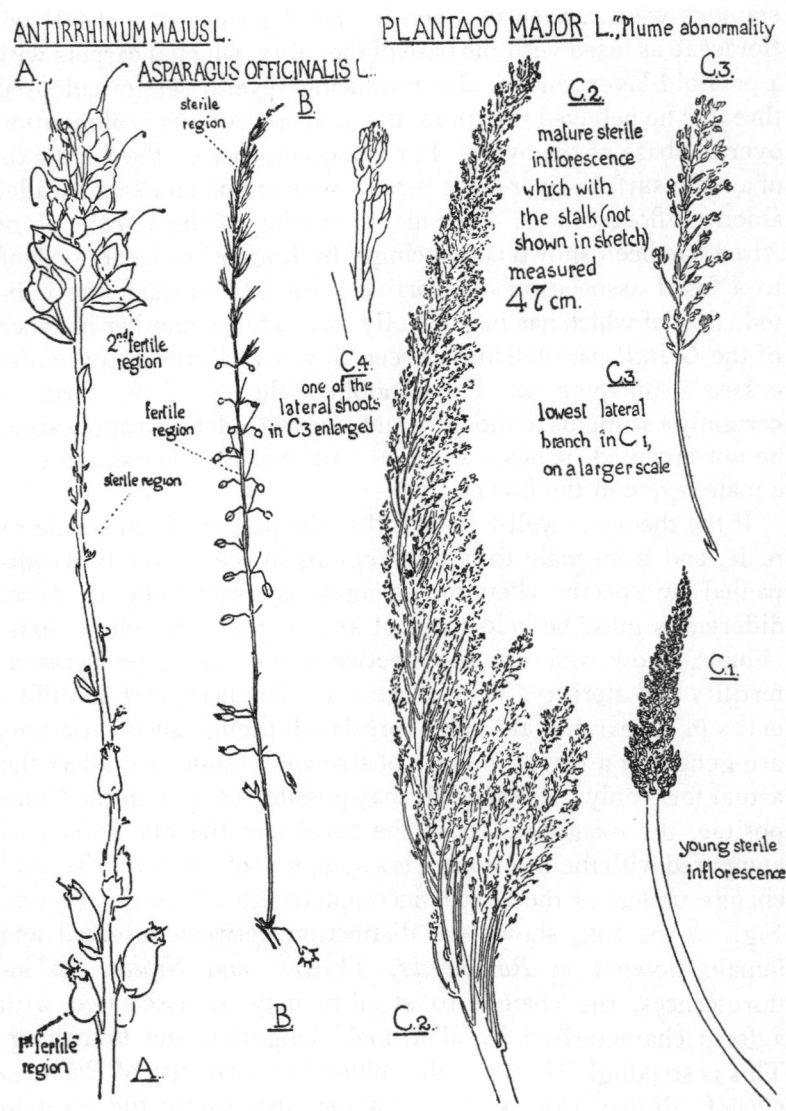

ANTIRRHINUM MAJUS L.

ASPARAGUS OFFICINALIS L.

PLANTAGO MAJOR L., "Plume abnormality

A

B

sterile region

2nd fertile region

fertile region

sterile region

1st fertile region

A

B

C4.
one of the lateral shoots in C3 enlarged

C2.
mature sterile inflorescence which with the stalk (not shown in sketch) measured 47 cm.

C3.
lowest lateral branch in C1, on a larger scale

C3.

C1.

young sterile inflorescence

C2.

Fig. 46. Zonal or complete sterility. A, B, C1, C2 (×½); C3, simplified; it had more than 20 branches; C3, C4 (enlarged).

194

from fertility back to a sterile phase; or when, as in *Antirrhinum* (Fig. 46, A), a similar retrogression is followed occasionally by a second period of fertility; it seems reasonable to suppose that these variations are associated with internal chemical oscillations. This is, however, a matter of hypothesis, but an example in which the existence of a relation between chemistry and form is left in no doubt, is afforded by the work of Raistrick, who showed that the classification of the genus *Aspergillus* on biochemical grounds coincided very closely with the existing morphological classification.[1] In this connexion the structural study of galls may offer a hopeful field, since these developments show what the plant *can* do under the influence of a stimulus, which is, presumably, in the main chemical.

The subtlety of the relations between chemistry and external form may well arouse hesitation as to the use of the word 'cause', which has often been too lightly applied in morphological argument in general. Goethe,[2] and various modern botanists, have held, for instance, that the disposition of the vascular system is the 'cause' of the external shape. No one would deny that the ground plan of the bundles and the form of the organ are intimately connected, but it cannot be said that either of these factors is the cause of the other, since their interrelationship is completely reciprocal. Certain localised changes may, it is true, be closely associated with the character of the vascular supply. We have seen, for instance, that there is a tendency to luxuriance, and sometimes to peloric development, in terminal flowers, which presumably receive the most direct tide of the food carried up the axis by the vascular system. In the leaves of pitcher plants, again, it appears possible to relate such peculiar features as the 'lids', and other appendages, to a localised richness of vascular supply. The development of the 'hood' in *Heliamphora nutans* Benth., and of the 'fishtail' appendage of *Darlingtonia californica* Torr., seem to be correlated with the convergence of main and lateral veins towards the leaf apex.[3]

When we turn from detailed instances to consider, on the broadest lines, the question of physico–chemical factors, we cannot but recognise that one such factor has been so universal as to

---

[1] Raistrick, H. (1932), p. 350.
[2] Goethe, J. W. von (1790), §20.   [3] Arber, A. (1941c).

affect practically the whole of the vegetable kingdom; this is the
development of a cell-wall encasing each unit of the plant body.[1]
The existence of cell-walls may have been due in the first place to
the power of photosynthesis, involving accumulation of excess
polysaccharides, which, as waste products, became deposited on
the surface of the protoplast. The plant, as Huxley wrote, "is an
animal confined in a wooden case, and Nature, like Sycorax,
holds thousands of 'delicate Ariels' imprisoned within every
Oak".[2] The fixity and immobility of the plant; colonial, repetitive
growth; the tree habit; and many other features, may be regarded
as bound up with the armoured character of the individual cell.
Movement, responsiveness, and all the possibilities which could
only be actualised in the absence of prison bars, were left to the
animal, which was less encumbered with the inhibiting products
of its own metabolism. We are concerned here with effects on
the largest scale, but even a minor alteration in the constitution
of the cell-wall may be associated with striking changes in the
external form. It has been shown that in the *compacta* mutants
of *Aquilegia vulgaris* L., various distinct characters are correlated
with precocious thickening of the walls; these effects include
uprightness of the flower; dwarfism; and a bushy, compact habit
of growth.[3]

When we review the factors which at different times have been
described in morphology as 'causal'—a few of which have been
touched upon in the present chapter—we cannot but feel that, in
general, a better name for them is 'conditioning' factors;[4] they
may perhaps be held to come into the category for which Aristotle
used the term 'conditional Necessity'.[5] Many years ago,
Coleridge wrote of the 'sophism' of "mistaking the *conditions*
of a thing for its *causes* and essence....The air I breathe, is the
condition of my life, not its cause".[6] As a botanical illustration
we may revert for a moment to the case of the anatropous ovule,

[1] Cf. Arber, A. (1928b); pp. 80–2; this paper was written in ignorance of
Huxley's anticipation of this idea.
[2] Huxley, T. H. (1853).
[3] Anderson, E. and Abbe, L. B. (1933), p. 383.
[4] For an illuminating discussion of the modes of action of lower unities, which
condition the modes of action of higher unities, see Russell, E. S. (1930), p. 185, etc.;
see also Thompson, D'Arcy W. (1942), p. 8.
[5] Peck, A. L. (1937), pp. 58, 59; 74, 75 [*De part. anim.* 639b; 642a].
[6] Coleridge, S. T. (1817), vol. I, p. 123.

in which development under pressure seems intimately related to the ultimate form.[1] On looking more closely into the matter, it becomes clear, however, that, whereas pressure in a crowded ovary must react very variously upon the ovules—according to their individual positions, and their space-relations to one another and to the ovary walls—the regular and consistent anatropous form, which finally results, is much more uniform than the pressures, which are held to have caused it, can possibly be. This difference seems to show that these pressures do not compel the curvature directly, as the pressure of the hand or tool enforces a certain shape upon a clay model; it appears that they merely convey an indication to the ovule, which awakens its potentiality for curvature of a particular kind. Something similar is suggested by other cases in which the effects of pressure during development continue to manifest themselves long after that pressure ceases to exist. In *Crambe maritima* L., for instance, the fruit has a narrow basal neck; there is little doubt that this narrowing is due to the pressure exerted on the gynaeceum in the young bud by the four broadly based and stipulate inner stamens. This state of compression must, however, come to an end quite early, but the peculiarity of form, which it has induced, remains a permanent feature—"As the twig's bent, so the tree's inclined." In the case of peloric flowers, again, it seems more reasonable to regard pressure as suggesting, but not enforcing, its behests. It is probable that release from pressure against the inflorescence axis, owing to a terminal position, should be regarded as conditioning rather than causing the exceptional regularity of the flower, for, as we have already noticed, flowers are sometimes peloric despite a lateral situation. After 200 years we must still echo the words of Rudberg: "What may be the cause of Linaria changing to Peloria has hitherto escaped us."[2]

Since, as Spinoza emphasised,[3] form (*figura*) is limitation, 'negative conditions' must be reckoned among the factors which control it. Both suppressions and inhibitions, and also their removal, would come into this category. Such factors may either

[1] See pp. 174, 175 (Fig. 37).
[2] Rudberg, D. (1744), in Linnaeus, C. (1749), *De Peloria*, p. 295, "Quaenam mutatae in Peloriam Linariae causa sit, nos adhuc fugit."
[3] [Spinoza, B. de] "B. D. S." (1677), letter to Jarig Jelles, 2 June 1674, p. 558; for translation see Wolf, A. (1928), p. 270.

act as a restraining influence, which prevents the full potentiality of the form from being reached, or they may act as releasing agents, which allow the form to reveal itself in its completeness. Such factors are no more causal than a break in a dyke, which may condition a flood, but can only be called its cause, if this word is denuded of much of its meaning. The idea that genic constitution is the *cause* of morphological features, is open to the same criticism; here the facts can be more accurately described on the theory that the genic outfit is *correlated* with morphological structure, and *conditions* it, though often in a negative sense.

Perhaps, as synthesising the considerations advanced in this chapter, we might compare the part played by physico-chemical influences in the development of form, with the part played by 'nurture' in the development of the human being, though frankly recognising that the analogy cannot be pressed beyond the preliminary point at which it serves to suggest a train of thought. In both cases the factors in question may be of the utmost importance, and may induce transformations of a striking kind; but such effects are, nevertheless, always confined within the unalterable boundaries set by inherent 'nature'. The modifications induced do not touch the inner core; in its essence the form remains unaffected.

CHAPTER XI

THE INTERPRETATION OF PLANT MORPHOLOGY

I N the preceding chapter attention has been called to some of the difficulties associated with the idea of 'cause' in morphology; this is a subject which now demands further study. When the general principles of biology are in question, we find, even to-day, that Aristotle[1] can often give us more fundamental help than any writer of a later date. In the course of his work he instituted a search into the nature of causation, which covers a wider field than most of the modern treatments of such themes, and in fact becomes an analysis of the whole range of attitudes open to the scientific investigator. Very briefly we may say that he distinguishes four primary causes:

(1) *The Material Cause*: the matter or substrate of the thing.

(2) *The Efficient Cause*: the source of motion, or the cause of change in the thing.

(3) *The Final Cause*: the purpose or end of the thing.

(4) *The Formal Cause*: the essence or essential nature of the thing.

This classification can be understood most easily in relation to the works of man, so, as a crudely simple illustration of what it means to think in such terms, we will consider a Roman road. This might be treated causally as follows:

(1) *The Material Cause*: the earth, stones, etc. used in the road making.

(2) *The Efficient Cause*: the forces set in motion by the labour employed in the construction.

---

[1] For an account of Aristotle's views, comprehensible to the biologist untrained in philosophy, see Stace, W. T. (1920). On the subject of the causes, see also Tredennick, H. (1933), pp. 16, 17; pp. 22, 23; pp. 416, 417, etc. [*Metaphys* I. iii. 1. 983*a*; I. iii. 11. 984*a*; VIII. iv. 4, 5. 1044*a*, *b*, etc.], and Mure, G. R. G. (1926), vol. I [*Anal. Post.* II. 11. 94*a* (Oxford trans.)]. Our present analysis of the causes does not follow Aristotle's order.

(3) *The Final Cause*: the purpose of facilitating travel from place to place, which is the reason for which the road is made.

(4) *The Formal Cause*: the idea of road-making which pre-existed in the Roman mind.

When studying organic form, we may consider these four causes as falling into two classes—the mechanical or physico-chemical causes (material + efficient causes), and the teleological causes (final + formal causes). As an example of the contrasting outlooks, which find their expression in these two types of causation, we may take the differing explanations given of the flattened and expanded character of the green leaf. On the one hand, dorsiventrality, regarded as an inevitable outcome of the partial-shoot structure of the leaf, may be held responsible, automatically, for the development of this extensive flat surface; for if a branch system (venation) is confined to one plane, it lends itself to webbing, through the non-separation of the regions served by each vein. This is a mechanical explanation; but the teleologist, observing the same features, would say that the dorsiventrality and expansion of the leaf have been developed in order to provide an effective assimilating, breathing, and transpiring organ.

Throughout the history of science, the varying views, which morphologists have taken as to the nature of form, are found to depend upon which of the Aristotelian causes, their own mentality, and the intellectual climate of the time, have led them to emphasise. At any one period we generally find that either the physico-chemical causes are stressed, and biologists lean to treating the organism as a machine; or the teleological causes are dominant, and the tendency is towards an interpretation in terms of purpose. Expressing the matter in the language of the child's storybook, we may say that the ascendancy has been gained by 'Madam How' at times when *causes*, in the popular sense, have been sought after, and by 'Lady Why', at times when the interest of mankind has turned rather to *reasons*. The seventeenth century saw the beginning of a period of insistence upon the Material and Efficient Causes—a period in which we are still living. It has, indeed, been maintained, especially by philosophers who take a somewhat externalised view of biology, that nothing has a right to the name of science, except the study of the Material and Efficient Causes, both of which are amenable to quantitative treatment.[1] A statement such as the following, from a book of the nineteen-thirties may serve as an example: "A plant is, in fact, the ideal embodiment of the organic machine, alive, but mechanical in what it does."[2] And even when their *reflections* are not thus restricted, researchers of to-day often confine themselves in *practice* to working with physico-chemical causes alone. D'Arcy Thompson, for instance, though recognising in theory that "like warp and woof, mechanism and teleology are interwoven together", yet wrote that the purpose of his book, *Growth and Form*,[3] "was to correlate with mathematical statement and physical law certain of the simpler outward phenomena of organic growth and structure or form, while all the while regarding the fabric of the organism, *ex hypothesi*, as a material and mechanical configuration". This mechanistic approach has the advantage of bringing the study of nature within the scope of controllable methods, and of transferring much of its difficulty to those manipulative and technical spheres in which the man of science is more happily at ease than in the world of thought. It has thus led to that enormous accumulation of facts, which is at once the pride and the burden of biology to-day. The physico-chemical mode of studying the animal or plant has, indeed, everything to recommend it, so long as its limitations are never forgotten, and it is realised that the application of this mode of study to the organism cannot, in the nature of things, reveal anything that lies outside physics and chemistry. This point has been made repeatedly by modern thinkers: Bertalanffy, for instance, stresses the fact that "the physico-chemical description of the vital processes does not exhaust them".[4] Even the most convinced mechanist, when he tries to describe the structure and ways of living things, is often found to slip into the language of teleology, thus showing that—at least subconsciously—he has not been able to rid himself of some faint acceptance of directiveness. Without this he would too much resemble the "mere physicist", whom the Persian mystic, al-Ghazālī, long ago

[1] Cf. for instance, Oakeshott, M. (1933), p. 222; for a criticism of such restrictions see Bertalanffy, L. von (1933), pp. 17, 18, etc., and Stocks, J. L. (1938), pp. 32 *et seq.*

[2] Savory, T. H. (1936), p. 27.

[3] Thompson, D'Arcy W. (1942); for continuations of various lines of thought initiated in this book see Clark, W. E. Le Gros, and Medawar, P. B. (1945).

[4] Bertalanffy, L. von (1933), p. 8.

XI. THE INTERPRETATION OF MORPHOLOGY

compared to "an ant which, as it crawls over a sheet of paper, observes black letters spreading over it, and refers the cause to the pen alone".[1] The exclusively mechanistic view of the organism, which was fostered by the physics and chemistry of the nineteenth century, is far narrower than the outlook of Aristotle, who, though assigning the fuller significance to the Final Cause, also recognised the importance of the Efficient Cause.[2] Theophrastus, Aristotle's successor, accepted the same duplex view. He realised the dependence of the plant upon physical conditions,[3] but he also, though with his customary caution, used teleological ways of thought; he noted, for instance, that the "ivy regularly puts forth roots from the shoots...by means of which it gets hold of trees and walls, as if these roots were made by nature on purpose".[4] Classical botany long maintained its hold, and Cesalpino, some 2000 years after Theophrastus, retained reliance upon both Efficient and Final Causes, though his terminology was rather different. He regards, for instance, the 'necessity' of the corolla and stamens as being to draw off the 'spiritus' from the turgid plant, whereas their 'purpose' is for the protection of the developing fruit.[5] Even in seventeenth-century England, when Aristotelianism was waning, and the New Philosophy, with its stress upon mechanism, was gaining ascendancy, there was co-existence of belief in mechanical causation and in teleology. This is revealed in the writings of Francis Bacon, who declares that both the physical cause and the final cause are "true and compatible";[6] and we find the same point of view in the speculations of the Cambridge Platonists. Ralph Cudworth, in 1678,[7] wrote that there is a mixture of directive causality[8] "together with *Mechanism*, which runs through the whole Corporeal Universe". We find, again,

---

[1] Smith, Margaret (1944), p. 69. (al-Ghazālī, A.D. 1058–1111.)
[2] Ogle, W. (1912), vol. v [*De part. anim.* I. 1. 639 b, 14 (Oxford trans.)].
[3] Hort, A. (1916), vol. I, pp. 18, 19 [I. ii. 4].
[4] Hort, A. (1916), vol. I, pp. 278, 279 [III, xviii. 10].
[5] Cesalpino, A. (1583), lib. I, cap. VII, p. 13.
[6] Bacon, F. (1605), bk. II, p. 30 (misnumbered 32 in the copy accessible to the writer).
[7] Cudworth, R. (1678), bk. I, chap. III, xxxvii, 3, p. 148.
[8] Cudworth's expression is "*Life* or *Plastick Nature*", but it is clear from other passages (e.g. bk. I, chap. III, xxxvii, 2, p. 147) that this phrase is used as synonymous with directive causality.

a respect for both these classes of causality in work of a very different kind—the observational writings of Nehemiah Grew. He sought earnestly for light from chemistry upon his problems, but he also expressed views that were frankly teleological. We may instance his description of the 2-ribbed bracteole of the spring crocus: "For having no *Empalement* [calyx], and starting up early out of the Mould, even before its *Green Leaves*, and that upon the first opening of the Spring; lest it should thus be quite starved, 'tis born swath'd up in a double Blanket, or with a pair of Sheets upon its Back."[1] Towards the end of the eighteenth century, Kant, looking at the matter from the standpoint of the philosopher, seems to recognise frankly the irreducible existence of the two types of cause. He noted that, in the animal body, many parts might be explained through laws of a merely mechanical kind ('nach bloss mechanischen Gesetzen'); but he added "Yet the cause that accumulates the appropriate material, modifies and fashions it, and deposits it in its proper place, must always be estimated teleologically ('immer teleologisch beurtheilt werden')".[2] Borrowing the oft-quoted expression, with which Kant illumined another occasion, we may say that mechanism without teleology is blind, while teleology without mechanism is empty.

Among British writers to-day, opinions as to the relation of teleology and mechanism are generally implicit rather than explicit: in the United States, on the other hand, the subject has been fully ventilated in recent years, but unfortunately much of this literature has received too little attention on this side of the Atlantic. Many references will be found in a book by R. S. Lillie,[3] which is a powerful stimulant to thought, whether or no the reader is able to accept the author's view of the relation of the 'physical' to the 'psychical'.

The problem that confronts us now is the search for a direction in which we may look for a synthesis of these two superficially irreconcilable modes of thought—the way of teleology, and the way of mechanism. Hitherto we have treated Aristotle's

[1] Grew, N. (1672), p. 135.
[2] Kant, I. (1790), *Kritik der teleologischen Urteilskraft*, quoted here from Kant, I. (1902–38), vol. v, 1908, p. 377; for translation see Meredith, J. C. (1928), p. 26.
[3] Lillie, R. S. (1945).

Final and Formal Causes together under the generalised term, teleology. It remains to see whether the situation becomes easier to grasp, if we distinguish these two causes as he did. When the Final Cause is considered in connexion with the living thing, it may be defined as the *extrinsic* directivity of the organism, that is to say, purposefulness directed to ends outside the organism itself. For the moment, we will set aside the Final Cause, and concentrate attention upon the Formal Cause, to which the biological approach is easier. To understand how it was that Aristotle came to think as he did of the Formal Cause, a notion which he first made explicit,[1] we must go back to Socrates and Plato, in whose work his philosophy was rooted. Socrates laid great stress upon *Concepts*, the nature of which can be indicated here only with undue simplification. As a crude botanical example, we may consider how man's mind deals with such a thing as a buttercup. An individual plant is first perceived and considered, and the result is the mental retention of a notion based upon the features of this individual. After the observation of a number of plants has given the intellect a series of images and symbols to work upon, a *general* notion of a thing called buttercup is formed, having the characteristics which have come to be recognised as common to the various individuals, which have been the objects of perception and thinking. This notion is *not* a composite picture, but an abstraction of the thought of 'butter-cupness' from everything which is irrelevant to this thought. Such general notions are called Universals or Concepts.[2] It is possible to hold that, to Socrates, these Concepts were merely part of the mechanism of thinking, and that he did not attribute to them any separate existence.[3] To Plato, on the other hand, Concepts, from which he developed his Ideas or Forms, bulked much larger in the scheme of things. He is generally believed to have regarded them as having an eternal being in a realm of their own, parallel to the world of sense-experience, but transcending it;[4] at times, however, he seems to have leaned to the immanence theory, according to which the Form existed, in some sense,

---

[1] Tredennick, H. (1933), pp. 48–51 [*Metaphys.* I. vii. 4. 988*a*, *b*].

[2] For a study of the status of Universals in the medieval period, when such problems were keenly debated, see Carré, M. H. (1946).

[3] Cf. Tredennick, H. (1935), pp. 196, 197 [*Metaphys.* XIII. iv. 5. 1078*b*].

[4] See especially Cornford, F. M. (1937), p. 192 [*Timaeus*, 51E–52].

within the individual. The latter view is, however, generally associated with Aristotle's name;[1] to him the Form thus became the being in its essence, or the Formal Cause. It may be regarded as representing that immanent and intrinsic purposefulness,[2] in which the whole is the cause of the parts; or, as Cudworth expressed it 2000 years later, "Art it self, acting immediately on the Matter, as an Inward Principle."[3] It must be realised that the Formal Cause differs in its nature from 'cause'—as that word is used in common parlance to denote something which *precedes* its effects—for Formal Causes *co-exist* with their effects, their priority being logical rather than temporal. The statement, that the Formal Cause is essentially immanent, may seem a contradiction in terms, since we are accustomed to associate cause with some degree of externality. It might be better to reject the word 'cause' in this connexion altogether, and to think of *reason* and *consequent*, rather than of *cause* and *effect*.[4]

Though he did not use Aristotle's phraseology, it was the Formal Cause which was in Sachs's mind when he spoke of "the something which always stands as a remainder in the investigation of organisms; which we recognise over-against non-living Nature, as wholly *sui generis*; for which we have not even exact terms; which can neither be measured nor weighed;...something metaphysical, if one means by this expression a factor lying behind or above physical, chemical, and mechanical processes".[5]

The most obvious example of the action of the Formal Cause within the plant is the process of development from the fertilised egg to maturity. Here the cells differentiate in due relation to one another, and co-operate in playing their individual and varied parts, until the mature structure, in all its elaboration, is finally reached. The process is entirely determinate. If the fertilised

---

[1] For a lucid discussion of the Forms, see Collingwood, R. G. (1945).

[2] For a convincing study of directiveness, especially in animals, see Russell, E. S. (1945).

[3] Cudworth, R. (1678), bk. I, chap. III, xxxvii, 8, p. 155.

[4] On the relation of 'cause' and 'reason', see Stace, W. T. (1924), chapter III.

[5] Pringsheim, E. G. (1932), p. 163; "es bleibt in der Erforschung der Organismen immer etwas übrig, was wir als ganz eigenartig gegenüber der leblosen Natur erkennen, wofür wir nicht einmal bestimmte Ausdrücke haben, ein Etwas, das sich weder messen noch wägen lässt...etwas Metaphysisches, wenn man unter diesem Ausdruck etwas bezeichnet, was hinter oder über den physikalischen, chemischen, mechanischen Vorgängen liegt." (This passage occurs in certain manuscript notes by Sachs, printed for the first time by Pringsheim.)

egg cell is normal, environmental influence may induce pathological changes, but nothing will make the sequence culminate in
any being other than an individual of the parent species, with its
characteristic morphology. In Sir Kenelme Digby's words,
written nearly three hundred years ago: "as long as nature
proceedeth in her regular course . . . so long (I say) it is impossible,
that any other thing in the World should grow (for example) out
of a little shrunk Akehorne, then a spread vast Oake; or out of
a single Bean, then that tall, green, tender Plant".[1] Moreover,
as, within the individual, the Formal Cause rises to full expression
when the organisation type is attained in the course of development, so, in a broader sense, its effect appears when one *Gestalt*
type[2] becomes manifest within different cycles of relationship.
When forms, which have a certain identity, though constructed
out of unlike elements (for instance a flower, and a flower-like
inflorescence) recur in different plants, we may regard this as the
outcome of Formal Causes working to parallel ends, but using
diverse materials. It is useful in this connexion to subdivide the
Formal Cause in accordance with the distinction which Goethe[3]
drew between *Bildung*, which is dynamic, and *Gestalt*, which is
the static view of the ultimate result reached by *Bildung*. We
may illustrate these relationships by comparison with a dance,
which we may regard as the *Bildung*, culminating in the grouping
of the individuals into a figure, which we may look upon as the
final *Gestalt*. The same ultimate figure might, however, be
reached as the end point of a wholly different series of movements;
but, by whatever paths the goal is attained, the Formal Cause,
leading up to this *Gestalt* grouping, consists in the intentions of
the individual dancers, each fulfilling some part of the scheme of
the designer of the ballet. This analogy, like all analogies between
the works of nature and of art, breaks down on the point that in
nature the directiveness *is* the life of the individual elements, and
is not imposed from the outside; it is as if the design of the ballet
were innate in the mind of each dancer. Returning to the analogy
of the Roman road, suggested earlier in this chapter, we may say
that—whereas the Formal Cause, in that case, was the idea of
road-making pre-existing in the Roman mind—if the road had

---

[1] Digby, Sir K. (1661), pp. 7, 8.   [2] See chapter IX.
[3] Troll, W. (1926), p. 116; Goethe, J. W. von, *Zur Morphologie*, 1807.

been a living organism, the Formal Cause would have existed within that organism, instead of in a mind external to it.

We may, then, conclude, that the organisation type and the *Gestalt* are both expressions of the Formal Cause. We have already considered the advantages of replacing the idea of the type by that of parallelism; we can now expand this suggestion by recognising that, in place of the *static* notion of the type, we must set the *dynamic* concept of *parallelism in action of the Formal Cause*—we must thus include the process (*Bildung*, parallel becoming[1]), as well as the result (*Gestalt*), in the Formal Cause.

In our present discussion of the Formal Cause, we have deliberately set aside the Final Cause, but to this we must now return. The Final Cause, as *extrinsic*, is commonly opposed to the Formal Cause, as *intrinsic*; but may not the two be synthesised? That this might be done was accepted even by Aristotle himself, who after defining the Formal Cause as the essence, and the Final Cause as the end, added that they might perhaps both be the same thing.[2] Going further, he identifies the causes thus fused, with the psyche, or first principle, of the living body.[3] This running-together of the Formal and Final Causes can only be justified if we no longer confine inherence to the Formal Cause, but recognise that both causes are essentially *immanent*—the Formal Cause being immanent in the organism, considered as a discrete individual, and the Final Cause being immanent in the organism, considered as a part of the Whole. If we take this view, many of the difficulties of teleology vanish, since most of these difficulties arise out of the fact that the purely extrinsic is unthinkable.

In discussing the partial-shoot theory of the leaf, we found help in Spinoza's doctrine that the organism's endeavour "to persevere in its own being is nothing but the actual essence of the thing itself".[4] This dictum is also closely relevant to our present attempt at interpretation, for we may identify Spinoza's *conatus*, or endeavour—which he holds to *be* life—with the Formal Cause, into which the Final Cause is merged. We thus reach the position

[1] See p. 161.
[2] Tredennick, H. (1933), pp. 416, 417 [*Metaphys.* VIII. iv. 5. 1044a].
[3] Hett, W. S. (1935), pp. 86–9 [*On the Soul*, II. iv. 415a, b].
[4] See pp. 76–78.

that the form (including in this term both the concepts of the organisation and the *Gestalt* types) is, actually, the inherent directiveness of the organism made manifest—Dante's "*virtù informante*".[1] It is possible to proceed yet a step further in the application of Spinoza's ideas. It may be recalled that he recognised no antithesis between mind and body, but regarded both as representing the same reality conceived, in the case of mind, under the attribute of thought, and, in the case of body, under the attribute of extension.[2] On corresponding lines we can free ourselves from the crippling effects of the antithesis between teleology and mechanism, by recognising that they are not inherently antagonistic, but that they represent the organism *under two different attributes*—the physico-chemical attribute and the attribute of inherent directiveness. Or again, we may adopt an attitude similar to that of Kant in his reconciliation of necessity and free will,[3] and say that physico-chemical thought deals with the phenomena, while morphological thought attempts to penetrate towards the thing-in-itself.

The division between mechanistic and teleological biology finds a parallel in the much-emphasised contrasts between logical analysis and metaphysics;[4] here, also, the antithesis is artificial. Seen justly, these two activities are not opposed to one another, but logical analysis occupies a special field of its own, within the wider domain of philosophy;[5] this field is comparable with the limited province of physico-chemical interpretations within the larger realm of synthetic biology.

A clear-cut difference between the teleological and the mechanistic modes of thought emerges when we consider how investigation proceeds in each case. There is no mystery about the attitude which the mechanistic approach involves, nor about the general nature of the technique to which the researcher is committed—a technique based essentially on analytic methods.

---

[1] *Paradiso*, canto VII, 133–41; cf. also the *virtute informativa* of *Purgatorio*, canto XXV, 41.

[2] [Spinoza, B. de] "B. D. S." (1677), *Ethices*, Pars II, prop. xxi, Schol. p. 65: "Mentem, et Corpus unum, et idem esse Individuum, quod jam sùb Cogitationis, jam sub Extensionis attributo concipitur."

[3] Kant, I. (1787), preface to edition 2 of *Kritik der reinen Vernunft* in Kant, I. (1902–38), vol. III, 1904, pp. 17–18; translated in Smith, N. Kemp (1933), pp. 27–8.

[4] See, for instance, Ayer, A. J. (1946).

[5] For a presentation of this synthetic view, cf. Barnes, W. H. F. (1947).

When on the other hand, we adopt the more distinctively morphological standpoint, from which form is treated as "the intuitively seized (anschaulich anfasste) totality of the organism, and, as such, unanalysable",[1] it is clear that a different mode of approach is inevitable. We have to find means, in the words of Goethe, "to master the Whole in some sort through the *Anschauung*".[2] The significance of *Anschauung* in such a context cannot be conveyed accurately in English without resorting to a lengthy phrase. As Goethe uses it, it may be held to signify the *intuitive knowledge gained through contemplation of the visible aspect*. It may be compared with 'imagination'—taking the word in the special sense in which it is used in Neo-Platonic doctrine to signify something analogous to "a mirror placed as a mean between thought and sense".[3] We may also regard *Anschauung* as combining the immediacy of *knowledge of*, with the mediate character of the *knowledge about*, which is reached by the analysable processes of discursive thought. It may, moreover, be compared with the word νοῦς (mind) as used by Aristotle to include both perception through the senses, and the intellectual faculties.[4]

The idea of *Anschauung* was closely connected in Goethe's mind with another concept—that of *Darstellung*. This word is as untranslatable as *Anschauung*; as used in morphology it may perhaps be rendered by the phrase, *interpretative portrayal*, or *internalised representation*. The morphologist has to aim at what the portrait-painter achieves when he adds intellectual insight to mastery of technique, so that his picture becomes a revelation of personality, as seen through, and expressed in, the external lineaments. Such considerations as these may seem remote from the day-to-day work of botanical research. If, however, we recur to such a history as that detailed earlier in this book, of the views which morphologists have held as to the nature of the leaf, and the modern outcome of these views in the partial-shoot theory, we shall realise that our present tentative solutions of such problems owe any value they may possess to the gradual refinement through the centuries of *Anschauung* and *Darstellung*.

---

[1] Troll, W. (1928*a*), p. 5.
[2] Troll, W. (1926), p. 115 [Goethe, J. W. von, *Zur Morphologie, Die Absicht eingeleitet*, 1807].
[3] Whittaker, T. (1918), p. 266.   [4] Hett, W. S. (1935), p. xi.

In the physico-chemical sciences interpretations are conveyed through the medium of words, supplemented by symbolism of a mathematical kind. In morphology, on the other hand, the second medium, added to words, is expression directed to the visual sense. The necessity for these two media is tacitly recognised by every botanist who illustrates his text with pictures of plants. As Turpin[1] wrote more than a hundred years ago, "La plume et le pinceau sont les deux principaux moyens dont nous puissions nous servir pour le signalement des êtres", and he added that, of the two instruments, pen and pencil, the pencil was perhaps "le plus significatif". There is, indeed, a certain correlation between artistic power and morphological insight. This may well be one of the reasons why the study of plant form was initiated and carried to so advanced a point in the classical period, for the visual capacity of the Greeks reached a peculiarly high level. It is not mere chance that in their language the words for 'knowing' (εἰδέναι) and 'seeing' (ἰδεῖν) came from the same stem, and that one term, θεωρία, was used both for scientific investigation and also for beholding.[2] The Greeks understood how to think with the mind's eye. The same visual element has been strong also in various later morphologists; Goethe is a conspicuous example. He preferred to call himself a *Naturschauer* (gazer into Nature) rather than a *Naturforscher* (investigator of Nature),[3] and he tells us that it was primarily through the eye that he apprehended the world.[4] He was at pains to learn both to draw and to engrave, and he took a lively interest in the history of plant delineation.[5] Sachs, again, whose insight led him to the truth that it was the shoot, rather than the stem or leaf, which should be regarded as the morphological unit, had much of the artist in him. His father was an engraver, and a portrait painter of no mean quality, and he himself had studied drawing from the antique.[6]

There is, indeed, much in the *facies* of the plant—the "general effect produced on the eye by the sum total of all the visible external characters"[7]—which cannot be expressed in words, but it can be portrayed by the artist, and can be used mentally by

[1] Turpin, P. J. F. (1820), p. 23.   [2] Strömberg, R. (1937), p. 177, footnote.
[3] Troll, W. (1926), p. 6.   [4] Troll, W. (1926), p. 76.
[5] See his article on *Blumenmalerei* (1818) reprinted in Goethe, J. W. von (Jubiläums-Ausgabe, n.d.), vol. xxxv, pp. 154–60.
[6] Pringsheim, E. G. (1932), pp. 12–13.   [7] Sprague, T. A. (1940), p. 449.

that combination of perceptive and intellectual activity, which results in the interpretative realisation of form. Sight is too often regarded as if it were merely a matter of physical sensation in which the mind is passive; but, when we say that we 'see' a form, this actually means that our inner powers have organised into a definite image the mosaic of stimuli which the eye has received.[1] This point is illustrated by the difference between a photograph and a drawing. The camera knows nothing, and can record merely that aspect of plant form which is actually presented to it, but the observer, who draws a plant, is not thus limited. He *knows* what the plant is like from other points of view, and the image which he organises is thus given a richness of content which a photograph inevitably lacks.[2] The passage from the sensory stimuli impressed upon the eye by a given form, to the imaginative realisation of that form in the mind, is a bridge between the senses and the intellect. While the *physico-chemical* study of form is achieved by means of conceptual reason, together with causal analysis, the *morphological* position is reached through combining conceptual reason, and thought which is visual and tactual. The first of these two modes of approach towards the understanding of form, involves reliance upon the Material and Efficient Causes alone, while the second invokes the reinstatement of Aristotle's Formal Cause,[3] with its corollary of thinking with the mind's eye. Though these two approaches seem, in our present limited view, to be parallel and disconnected, yet—since the four causes are all abstractions—the two paths must converge at last upon the ultimate concrete synthesis of the work of the eye and the intellect. There is a tale[4]—as significant in essence as it is historically baseless—that Avicenna, the philosopher, and Abú Sa'íd, the mystic, once met. After they had conversed for three days and three nights, the man of analytic thought rounded their discourse in the words, "All that I know, he sees", while the man of vision said," All that I see, he knows."

---

[1] For this, as the Socratic view expressed by Plato, see Taylor, A. E. (1929), pp. 338–9, and for a modern application of it, Russell, E. S. (1932), p. 168.

[2] Cf. Wilmott, A. J. (1945).

[3] The necessity of a return in the study of nature to the Aristotelian Formal Cause has been stressed from the philosophic standpoint in Stocks, J. L. (1939); see p. 80.

[4] Nicholson, R. A. (1921), p. 42.

14-2

LIST OF BOOKS & MEMOIRS CITED

WITH PAGE REFERENCES TO THE TEXT

AASE, H. C. (1915). Vascular anatomy of the megasporophyll of conifers. *Bot. Gaz.* vol. LX, pp. 277–313. [p. 127]

ABBE, L. B.; *see* ANDERSON, E. and ABBE, L. B. (1933).

ALBERTUS MAGNUS; *see* MEYER, E. H. F. and JESSEN, K. F. W. (1867).

ALBRECHT, J. S.; *see* JUNG, J. (1747).

ALLEN, G. S. (1947). Embryogeny and the development of the apical meristems of *Pseudotsuga*. III. Development of the apical meristems. *Amer. Journ. Bot.* vol. XXXIV, pp. 204–11. [p. 135]

ALMEIDA, F. D' (1824). Notice sur la vie et les travaux de M. Corréa de Serra. *Mém. Mus. Hist. Nat.* vol. XI, Paris, pp. 215–29. [p. 60]

ANDERSON, A. and ABBE, L. B. (1933). A comparative anatomical study of a mutant *Aquilegia*. *Amer. Nat.* vol. LXVII, pp. 380–4. [p. 196]

ANDREWS, H. N. (1941). *Dichophyllum Moorei* and certain associated seeds. *Ann. Mo. Bot. Gdn*, vol. XXVIII, pp. 375–84. [p. 75]

ANON. (1526). *The Grete Herball*. London: Peter Treveris. [p. 153]

ARBER, A. (1918). The phyllode theory of the monocotyledonous leaf. *Ann. Bot.* vol. XXXII, pp. 465–501. [pp. 85, 108]

ARBER, A. (1921). The leaf structure of the Iridaceae, considered in relation to the phyllode theory. *Ann. Bot.* vol. XXXV, pp. 301–36. [pp. 85, 102, 107]

ARBER, A. (1922a). Leaves of the Farinosae. *Bot. Gaz.* vol. LXXIV, pp. 80–94. [p. 99]

ARBER, A. (1922b). On the nature of the 'blade' in certain monocotyledonous leaves. *Ann. Bot.* vol. XXXVI, pp. 329–51. [p. 85]

ARBER, A. (1924a). *Danae, Ruscus*, and *Semele*: A Morphological Study. *Ann. Bot.* vol. XXXVIII, pp. 229–60. [pp. 97, 109]

ARBER, A. (1924b). *Myrsiphyllum* and *Asparagus*: A Morphological Study. *Ann. Bot.* vol. XXXVIII, pp. 635–59. [p. 97]

ARBER, A. (1925). *Monocotyledons: A Morphological Study*. Cambridge. [pp. 30, 85, 99, 102, 107, 108, 160]

ARBER, A. (1927). Studies in the Gramineae. III. Outgrowths of the reproductive shoot, and their bearing on the significance of lodicule and epiblast. *Ann. Bot.* vol. XLI, pp. 473–88. [p. 94]

ARBER, A. (1928a). Studies in the Gramineae. V. 1. On *Luziola* and *Dactylis*. 2. On *Lygeum* and *Nardus*. *Ann. Bot.* vol. XLII, pp. 391–407. [p. 94]

ARBER, A. (1928b). The tree habit in angiosperms. *New Phytol.* vol. XXVII, pp. 69–84. [p. 196]

ARBER, A. (1929a). Studies in the Gramineae. VI. 1. *Streptochaeta*. 2. *Anomochloa*. 3. *Ichnanthus*. *Ann. Bot.* vol. XLIII, pp. 35–53. [pp. 56, 94]

ARBER, A. (1929b). Studies in the Gramineae. VIII. On the organization of the flower in the bamboo. *Ann. Bot.* vol. XLIII, pp. 765–81. [p. 56]

ARBER, A. (1930). Root and shoot in the angiosperms. *New Phytol.* vol. XXIX, pp. 297–315. [pp. 71, 92, 166]

ARBER, A. (1931a). Studies in floral morphology. II. On some normal and abnormal crucifers: with a discussion on teratology and atavism. *New Phytol.* vol. XXX, pp. 172–203. [pp. 5, 92, 99, 141, 173, 180]

ARBER, A. (1931b). Studies in floral morphology. III. On the Fumarioideae, with special reference to the androecium. *New Phytol.* vol. XXX, pp. 317–54. [pp. 94, 127]

ARBER, A. (1932). Studies in floral morphology. IV. On the Hypecoideae, with special reference to the androecium. *New Phytol.* vol. XXXI, pp. 145–73. [pp. 91, 127]

ARBER, A. (1934). *The Gramineae: A Study of Cereal, Bamboo, and Grass.* Cambridge. [pp. 6, 47, 56, 87, 149, 160, 173]

ARBER, A. (1935). The 'needles' of *Asparagus*, with special reference to *A. Sprengeri* Reg. *Ann. Bot.* vol. XLIX, pp. 337–44. [p. 97]

ARBER, A. (1936). Studies in flower structure. II. On the vascular supply to the nectary of *Ranunculus*. *Ann. Bot.* vol. L, pp. 305–19. [p. 82]

ARBER, A. (1937a). The interpretation of the flower: a study of some aspects of morphological thought. *Biol. Rev.* vol. XII, pp. 157–84. [pp. 2, 47, 59, 124, 144, 151]

ARBER, A. (1937b). Studies in flower structure. III. On the 'corona' and androecium in certain Amaryllidaceae. *Ann. Bot.*, N.S., vol. I, pp. 293–304. [pp. 82, 104]

ARBER, A. (1938a). *Herbals: A Chapter in the History of Botany, 1470–1670.* Edition 2. Cambridge. [pp. 32, 123]

ARBER, A. (1938b). Studies in flower structure. IV. On the gynaeceum of *Papaver* and related genera. *Ann. Bot.*, N.S., vol. II, pp. 649–64. [p. 104]

ARBER, A. (1940). Studies in flower structure. VI. On the residual vascular tissue in the apices of reproductive shoots, with special reference to *Lilaea* and *Amherstia*. *Ann. Bot.*, N.S., vol. IV, pp. 617–27. [pp. 56, 95, 151]

ARBER, A. (1941a). The interpretation of leaf and root in the angiosperms. *Biol. Rev.* vol. XVI, pp. 81–105. [pp. 70, 131]

ARBER, A. (1941b). Nehemiah Grew and Marcello Malpighi. *Proc. Linn. Soc. Lond.* Session 153, pt. 2, Nov. 21, pp. 218–38. [pp. 16, 38]

ARBER, A. (1941c). On the morphology of the pitcher-leaves in *Heliamphora, Sarracenia, Darlingtonia, Cephalotus*, and *Nepenthes*. *Ann. Bot.*, N.S., vol. V, pp. 563–78. [pp. 157, 195]

ARBER, A. (1942). Studies in flower structure. VII. On the gynaeceum of *Reseda*, with a consideration of paracarpy. *Ann. Bot.*, N.S., vol. VI, pp. 44–8. [p. 58]

ARBER, A. (1943). Spinoza and Boethius. *Isis*, vol. XXXIV, pp. 399–403. [p. 77]

ARBER, A. (1946a). Goethe's Botany. *Chronica Botanica* (Waltham, Mass. U.S.A.), vol. X, no. 2, pp. 63–126. (Separate publication.) [p. 40]

ARBER, A. (1946b). Analogy in the History of Science, pp. 221–33. In *Studies and Essays in the History of Science and Learning, presented to George Sarton*. Ed. by M. F. Ashley Montagu. New York. [pp. 13, 158]

ARBER, A. (forthcoming). From Medieval Herbalism to the Birth of Modern Botany. In *Essays presented to Charles Singer*. Ed. by E. Ashworth Underwood. [p. 37]

ARBER, A.; *see also* ROBERTSON, A. (1906).

Aristotle (translations); *see* DENGLER, R. E. (1927), FORSTER, E. S. (1913), HETT, W. S. (1935, 1936), MURE, G. R. G. (1926), OGLE, W. (1897, 1912), PECK, A. L. (1937), THOMPSON, D'ARCY W. (1910), TREDENNICK, H. (1933, 1935).

ARNAL, C. (1945). Recherches Morphologiques et Physiologiques sur la Fleur des Violacées. Thèse: Docteur ès sci. nat., Dijon. [p. 47]

ASHBY, E. (1937). The determination of size in plants. *Proc. Linn. Soc. Lond.* Session 149 (1936–7), pt. 2, pp. 59–65. [p. 190]

AUGUSTINE, SAINT, *see* HEALEY, J. (1610), and HEALEY, J. and TASKER, R. V. G. (1945).

AVERY, G. S. (1933). Structure and development of the tobacco leaf. *Amer. Journ. Bot.* vol. xx, pp. 565–92. [pp. 79, 171]

AYER, A. J. (1946). *Language, Truth and Logic*. London. [p. 208]

BACON, F. (1605). *The Twoo Bookes....Of the proficience and advancement of Learning, divine and humane*. London. [p. 202]

BACON, F. (1620). *Instauratio magna*. London. [pp. 5, 27]

BACON, F. (1631). *Sylva sylvarum*. Edition 3. London. [pp. 12, 27]

BACON, F.; *see* KITCHIN, G. W. (1855).

BAILEY, I. W.; *see* SINNOTT, E. W. and BAILEY, I. W. (1914).

BALFOUR, I. B.; *see* SACHS, J. VON (1890).

BARNES, B. (1933). Notes on *Lathyrus Aphaca* L. *Journ. Bot.* vol. LXX, pp. 10–15, 25–33. [p. 99]

BARNES, W. H. F. (1947). Is philosophy possible? A study of logical positivism. *Philosophy*, vol. XXII, pp. 25–48. [p. 208]

BARY, A. DE (1884). *Comparative Anatomy of the Vegetative Organs of the Phanerogams and Ferns*. Trans. by F. O. Bower and D. H. Scott. Oxford. [pp. 91, 126, 127]

BATESON, W. (1916). Root-cuttings, chimaeras and 'Sports'. I. *Journ. Gen.* vol. VI (1916–17), pp. 75–80. [p. 134]

BATESON, W. (1921). Root-cuttings and chimaeras. II. *Journ. Gen.* vol. XI, pp. 91–7. [p. 134]

BAYLE, P. (1820). *Dictionnaire Historique et Critique*, vol. V, new ed. Paris. [p. 28]

BENNETT, A. W.; *see* SACHS, J. VON (1875).

BERTALANFFY, L. VON (1933). *Modern Theories of Development: an introduction to theoretical biology*. Trans. by J. H. Woodger. Oxford. [p. 201]

BIDNEY, D. (1940). *The Psychology and Ethics of Spinoza*. New Haven: Yale University Press. [p. 77]

BISWAS, K. (1935). Observations on the systematic position of *Ficus Krishnae*. *Current Science*, vol. III, pp. 424–5. [p. 110, legend of Fig. 15]

BLASER, H. W. & EINSET, J. (1948). Leaf development in six periclinal chromosomal chimeras of apple varieties. *Amer. Journ. Bot.* vol. XXXV, pp. 473–82. [p. 72]

BOETHIUS, A. M. T. S. (1609). *Five Bookes Of Philosophical Comfort.... Newly Translated out of Latine* [by "I.T."]. London. [pp. 12, 77]

BOETHIUS, A. M. T. S.; see STEWART, H. F. and RAND, E. K. (1918).

BOWER, F. O. (1923). The relation of size to the elaboration of form and structure of the vascular tracts in primitive plants. *Proc. Roy. Soc. Edinb.* vol. XLIII (1922–3), pp. 117–26. [p. 190]

BOWER, F. O. (1930). *Size and Form in Land Plants.* London. [p. 190]

BOWER, F. O. (1931). Size and form in plants. *Rep. Brit. Ass. Adv. Sci.*, 98th Meeting, Bristol, 1931 for 1930, Pres. Add., pp. 1–14. [p. 190]

BRAUN, A. (1853). Das Individuum der Pflanze. *Abh. Kgl. Akad. Wissensch., Berlin*, pp. 21–122. [p. 128]

BRÉHIER, E. (1927). *Plotin. Ennéades.* vol. IV. Paris. [p. 158]

BRIEGER, F. G. (1935). The developmental mechanics of normal and abnormal flowers in *Primula*. *Proc. Linn. Soc. Lond.* Session 1934–5, pt. IV, pp. 126–9. [pp. 174, 191]

BRIQUET, J. (1935–6). Les caractères de la disymétrie et de l'hétérophyllie foliaire chez les Méliacées à feuilles composées. *Boissiera (Supp. de Candollea), Mém. Inst. Nat. Genev.* vol. XXIV, 126 pp. [p. 120]

BROOKS, F. T. and STILES, W. (1910). The structure of *Podocarpus spinulosus* (Smith) R.Br. *Ann. Bot.* vol. XXIV, pp. 305- 18. [pp. 127, 129]

BROOKS, R. M. (1940). Comparative histogenesis of vegetative and floral apices in *Amygdalus communis*, with special reference to the carpel. *Hilgardia*, vol. XIII, pp. 249–99. [p. 124]

BROWN, R. (1810). *Prodromus Florae Novae Hollandiae.* Vol. I. (No further volumes published.) London. [p. 66]

BROWN, R. (1822). An account of a new genus of plants, named *Rafflesia*. *Trans. Linn. Soc. Lond.* vol. XIII, pp. 201–34. [p. 68]

BROWN, R. (1840). On the relative position of the divisions of stigma and parietal placentae (reprinted from *Plantae Javanicae Rariores*, pt. II, pp. 107–12) in *Miscellaneous Botanical Works of R.B.*, Ray Soc., Lond., vol. I, 1846, pp. 553–63. [p. 171]

BROWNE, Sir T. (1658): *Hydriotaphia... Together with The Garden of Cyrus.* London. (Facsimile in the Noel Douglas Replicas, London, 1927.) [p. 50]

CABOT, R. C. (1904). *Clinical Examination of the Blood.* Edition 5. London. [p. 11]

CALMAN, W. T. (1930). The Taxonomic Outlook in Zoology. Pres. Add. to Sect. D, Zoology. *Rep. Brit. Ass. Adv. Sci.*, Bristol, 1931 (for 1930), pp. 83–91. [p. 65]

CANDOLLE, A. CASIMIR P. DE (1868). Théorie de la feuille. *Arch. Sci. Phys. Nat. Genève*, N.P., vol. XXXII, pp. 31–64. [pp. 74, 109]

CANDOLLE, A. CASIMIR P. DE (1899a). Sur les feuilles peltées. *Bull. trav. Soc. bot. Genève*, Années 1898–9, no. 9, 51 pp. [pp. 109, 111]

CANDOLLE, A. CASIMIR P. DE (1899b). On the peltation of leaves. *Rep. Brit. Ass. Adv. Sci.*, Bristol, 1899 (for 1898), p. 1065. [p. 109]

LIST OF BOOKS AND MEMOIRS CITED

CANDOLLE, A. P. DE (1813). *Théorie Élémentaire de la Botanique*. Paris. [pp. 46, 60, 173, 176, 177]

CANDOLLE, A. P. DE (1827). *Organographie végétale*. 2 vols. Paris. [pp. 5, 30, 41, 46, 61, 70, 84, 166, 176, 177, 191]

CANDOLLE, A. P. DE (1862). *Mémoires et Souvenirs*. Genève. [pp. 46, 60]

CARRÉ, M. H. (1946). *Realists and Nominalists*. Oxford. [p. 204]

ČELAKOVSKÝ, L. J. (1878). *Teratologische Beiträge zur morphologischen Deutung des Staubgefässes. Prings. Jahrb. wiss. Bot.* vol. XI, pp. 124–74. [p. 65]

CESALPINO (CAESALPINUS), A. (1583). *De plantis libri XVI*. Florence. [pp. 21, 28–32, 202]

CHAMBERLAIN, C. J. (1935). *Gymnosperms: Structure and Evolution*. Chicago. [p. 128]

CHAMBERLAIN, C. J.; see also COULTER, J. M. and CHAMBERLAIN, C. J. (1917).

CHAMISSO, A. DE and SCHLECHTENDAL, D. F. L. VON (1826). Plantis speculatoria romanzoffiana observatis (Umbelliferae). *Linnaea*, vol. I, pp. 333–401. [p. 168]

CHODAT, R. (1920). Ombellifères (in *La Végétation du Paraguay*, CHODAT, R. and VISCHER, W.), *Bull. Soc. Bot. Genève*, ser. 2, vol. XII, pp. 25–54. [p. 159]

CHRIST (CHRIST-SOCIN), H. (1897). *Die Farnkräuter der Erde*. Jena. [p. 116]

CHRYSLER, M. A.; see MCINTYRE, W. G. and CHRYSLER, M. A. (1943).

CLARK, W. E. LE GROS and MEDAWAR, P. B. (1945). *Essays on Growth and Form, presented to D'Arcy Wentworth Thompson*. Oxford. [p. 201]

COLERIDGE, S. T. (1817). *Biographia Literaria*. 2 vols. London. [p. 196]

COLLINGWOOD, R. G. (1945). *The Idea of Nature*. Oxford. [p. 205]

CORNFORD, F. M. (1937). *Plato's Cosmology. The Timaeus . . . translated with a running commentary by F. M. C.* London. [pp. 2, 9, 12, 26, 65, 204]

CORRÉA DA SERRA, J. F. (1796). On the fructification of the submersed Algae. *Phil. Trans. Roy. Soc.*, Lond. (no volume number), pp. 494–505. [p. 60]

CORRÉA DA SERRA, J. F. (1805). Observations sur la famille des orangers et sur les limites qui la circumscrivent. *Ann. Mus. Hist. Nat.* vol. VI, Paris, pp. 376–87. [p. 174]

CORRÉA DA SERRA, J. F. (1806). Observations carpologiques. *Ann. Mus. Hist. Nat.* vol. VIII, pp. 59–76, 389–400. [p. 60]

CORRÉA DA SERRA, F. J. (1807). Vues carpologiques. *Ann. Mus. Hist. Nat.* vol. X, pp. 151–62. [p. 160]

COULTER, J. M. and CHAMBERLAIN, C. J. (1917). *Morphology of Gymnosperms*. Edition 2. Chicago. [p. 128]

CROSS, G. L. (1938). A comparative histogenetic study of the bud scales and foliage leaves of *Viburnum opulus*. *Amer. Journ. Bot.* vol. XXV, pp. 246–58. [p. 86]

CROW, W. B. (1929). *Contributions to the Principles of Morphology*. London. [p. 3]

CUDWORTH, R. (1678). *The True Intellectual System of the Universe*. London. [pp. 202, 205]

LIST OF BOOKS AND MEMOIRS CITED

DARWIN, C. (1859). *On the Origin of Species.* London. [p. 160]

DELPINO, F. (1869). Ulteriori osservazioni...sulla Dicogamia. *Atti Soc. Ital. Sci. Nat.* vol. xii, pp. 21–141, 179–233. [pp. 155, 182]

DELPINO, F. (1889 a). Valore morfologico della squama ovulifera delle Abietinee e di altre Conifere. *Malpighia,* vol. iii, ann. iii, pp. 97–100. [p. 129]

DELPINO, F. (1889 b). Applicazione di nuovi criterii per la classificazione della piante. Mem. ii. *Mem. R. Accad. Bologna,* ser. iv, T.x, pp. 43–75. [p. 129]

DENGLER, R. E. (1927). *Theophrastus: De Causis Plantarum.* Book i. Philadelphia. [pp. 20–23, 25]

DICKIE, G.; *see* M'COSH, J. and DICKIE, G. (1856).

DIGBY, Sir K. (1661). *A Discourse Concerning the Vegetation of Plants.* Spoken by Sir K. D. at Gresham College, on the 23. of January, 1660. London. [p. 206]

DINGLER, H. (1885). *Die Flachsprosse der Phanerogamen.* Heft i. *Phyllanthus, sect. Xylophylla.* München. [p. 89]

DIOSCORIDES; *see* GUNTHER, R. T. (1934).

DITTMER, H. J. (1937). A quantitative study of the roots and root hairs of a winter rye plant (*Secale cereale*). *Amer. Journ. Bot.* vol. xxiv pp. 417–20. [p. 132]

DIVER, C. (1940). The problem of closely related species living in the same area. In *The New Systematics,* edited by J. S. Huxley, Oxford, pp. 303–28. [p. 4]

DODOENS, R. *see* LYTE, H. (1578).

DOMIN, K. (1909). Morphologische und phylogenetische Studien über die Familie der Umbelliferen. *Bull. Int. Acad. Prague,* Teil i, Année xiii (1909 for 1908), pp. 108–53; Teil ii, Année xiv (1909), pp. 1–52. [p. 139]

DORMER, K. J. (1945). An investigation of the taxonomic value of shoot structure in angiosperms with especial reference to Leguminosae. *Ann. Bot.,* N.S., vol. ix, pp. 141–53. [p. 127]

DRESSER, C. (1859). *Unity in Variety, as Deduced from the Vegetable Kingdom: being an attempt at developing that oneness which is discoverable in the habits, mode of growth, and principle of construction of all plants.* London. [pp. 74, 144]

DUCHARTRE, P. (1881). Note sur des feuilles ramifères de chou. *Bull. Soc. Bot. France,* vol. xxviii (ser. 2, vol. iii), pp. 256–64. [p. 105]

DUFRENOY, J. (1918). Pine needles, their significance and history. *Bot. Gaz.* vol. lxvi, pp. 439–54. [p. 75]

DYER, W. T. THISELTON; *see* SACHS, J. VON (1875).

EAMES, A. J. (1913). The morphology of *Agathis australis. Ann. Bot.* vol. xxvii, pp. 1–38. [p. 127]

ECKARDT, T. (1937). Untersuchungen über Morphologie, Entwicklungsgeschichte und systematische Bedeutung des pseudomonomeren Gynoeceums. *Nova Acta Leopoldina,* N.F., vol. v, no. 26, 112 pp. [p. 58]

ECKERMANN, J. P. (1836, 1848). *Gespräche mit Goethe in den letzten Jahren seines Lebens* (1823–32). Vols. i and ii, Leipzig, 1836; vol. iii, Magdeburg, 1848. [p. 41]

217

EICHLER, A. W. (1863). Excursus morphologicus de formatione florum Gymnospermarum. *Ann. sci. nat.* ser. IV, vol. XIX, pp. 257–85. [p. 130]

EICHLER, A. W. (1875, 1878). *Blüthendiagramme.* Leipzig. [p. 171]

EICHLER, A. W. (1881). Ueber die weiblichen Blüthen der Coniferen. *Mbr. preuss. Akad. Wiss.,* Berlin. Jahrg. 1882 for 1881, pp. 1020–49. [p. 130]

EICHLER, A. W. (1889). Coniferae. In *Die natürlichen Pflanzenfamilien* (ENGLER, A. and PRANTL, K.), II, 1, pp. 28–116. [p. 91]

EINSET, J., *see* BLASSER, H. W. & EINSET, J. (1948).

ENGLER, A. (1884). Beiträge zur Kenntniss der Araceae. V. *Engler's Bot. Jahrb.* vol. V, pp. 287–336. [p. 193]

ENGLER, A. (1897). *Nachtrag zu Teil II–IV, Die natürlichen Pflanzen-familien.* Leipzig. [p. 170]

ENGLER, A.; *see* WOLFF, H. (1913).

ESAU, K. (1943). Origin and development of primary vascular tissues in seed plants. *Bot. Rev.* vol. IX, pp. 125–206. [p. 162]

FORSTER, E. S. (1913). (In *The Works of Aristotle translated into English.* Vol. VI.) *De Plantis.* Oxford. [pp. 24, 76]

FOSTER, A. S. (1929). Investigations on the morphology and comparative history of development of foliar organs. I. The foliage leaves and cataphyllary structures in the horsechestnut (*Aesculus Hippocastanum* L.). *Amer. Journ. Bot.* vol. XVI, pp. 441–74, 475–501. [p. 118]

FOSTER, A. S. (1936a). Leaf differentiation in angiosperms. *Bot. Rev.* vol. II, pp. 349–72. [p. 108]

FOSTER, A. S. (1936b). A neglected monograph on foliar histogenesis. *Madroño,* vol. III, pp. 321–5. [p. 141]

FOSTER, A. S. (1939). Problems of structure, growth and evolution in the shoot apex of seed plants. *Bot. Rev.* vol. V, no. 8, pp. 454–70. [p. 162]

FRIES, R. E. (1909). Zur Kenntnis der Blattmorphologie der Bauhinien und verwandter Gattungen. *Arkiv. Bot.* Bd. VIII, No. 10, 16 pp. [p. 99]

GAISBERG, E. VON (1922). Zur Deutung der Monokotylenblätter a Phyllodien, unter besonderer Berücksichtigung der Arbeit von A. Arber; 'The Phyllode Theory of the Monocotyledonous Leaf, with Special Reference to Anatomical Evidence.' *Flora,* N.F., vol. XV (C.R.115), pp. 177–190. [p. 85]

GARNSEY, H. E. F.; *see* SACHS, J. VON (1890).

GEGENBAUR, C. (1859 and 1870). *Grundzüge der Vergleichenden Anatomie.* Editions I and II. Leipzig. [p. 63]

GEOFFROY SAINT-HILAIRE, I. (1832–6). *Histoire générale et particulière des anomalies de l'organisation chez l'homme et les animaux.* 3 vols. Paris. [p. 5]

GERARD, J. (1633). *The Herball or Generall Historie of Plantes....Very much Enlarged and Amended by Thomas Johnson.* London. [p. 9]

GEYLER, H. T. (1867). Ueber den Gefässbündelverlauf in den Laubblatt-regionen der Coniferen. *Prings. Jahrb. wiss. Bot.* vol. VI, 1867–8, pp. 55–208. [p. 127]

GEYLER, H. T. (1881). Einige Bemerkungen über *Phyllocladus. Abh. Senckenb. Naturf. Ges.* vol. XII, pp. 209–14 and 216. [p. 127]

LIST OF BOOKS AND MEMOIRS CITED

GILLETT, E. C.; see PRIESTLEY, J. H., SCOTT, L. I. and GILLETT, E. C. (1935).

GLÜCK, H. (1911). *Biologische und morphologische Untersuchungen über Wasser- und Sumpfgewächse*. Pt. III. Jena. [p. 116]

GLÜCK, H. (1919). *Blatt- und blütenmorphologische Studien*. Jena. [p. 50, 141]

GODRON, D. A. (1864 a). Mémoire sur les Fumariées à fleurs irrégulières et sur la cause de leur irrégularité. *Ann. sci. nat. (Bot.)*, ser. v, vol. II, pp. 272–80. [p. 173]

GODRON, D. A. (1864 b). Mémoire sur l'inflorescence et les fleurs des Crucifères. *Ann. sci. nat. (Bot.)*, ser. v, vol. II, pp. 281–305. [p. 173]

GOEBEL, K. VON (1933). *Organographie der Pflanzen*. Edition 3. Pt. III, *Samenpflanzen*. Jena. [pp. 50, 166]

GOETHE, J. W. VON (1790). *Versuch die Metamorphose der Pflanzen zu erklären*. Gotha. [pp. 5, 40–45, 73, 173, 191, 195]

GOETHE, J. W. VON (1831). *Versuch über die Metamorphose der Pflanzen. Übersetzt von F. Soret, nebst geschichtlichen Nachträgen*. Stuttgart. [p. 41]

GOETHE, J. W. VON (1887 etc.) *Werke herausgegeben im Auftrage der Grossherzogin Sophie von Sachsen*. Weimar. [pp. 40, 73, 132]

GOETHE, J. W. VON (1902, etc.). *Sämtliche Werke*. Jubiläums-Ausgabe. Stuttgart and Berlin. Vol. XXXV, *Schriften zur Kunst*, pp. 154–60. [p. 210]

GOETHE, J. W. VON (works and translations); see ECKERMANN, J. P. (1836, 1848), TROLL, W. (1926), ARBER, A. (1946 a).

GOODYER, J.; see GUNTHER, R. T. (1934).

GRAHAM, R. J. D. and STEWART, L. B. (1929). Vegetative propagation. Leaf cuttings in gymnosperms. *Trans. Bot. Soc. Edinb.* vol. XXX, pt. II, pp. 67–9. [p. 107]

GRAY, A. (1858). *Structural Botany*. Edition 5. New York. [p. 68]

GRAY, A. (1887). *Structural Botany*. Edition 6. London. [p. 68]

GREENE, E. L. (1909). Landmarks of Botanical History. Part I. Prior to 1562 A.D. *Smithsonian Misc. Coll.* No. 1870. Part of vol. LIV. Washington. [p. 18]

GRÉGOIRE, V. (1938). La morphogénèse et l'autonomie morphologique de l'appareil floral. I. Le Carpelle. *La Cellule*, vol. XLVII, pp. 287–452. [p. 47]

GREGORY, F. G.; see PURVIS, O. N. and GREGORY, F. G. (1937).

GREW, N. (1672). *The Anatomy of Vegetables Begun. With a General Account of Vegetation Founded thereon*. London. [pp. 37–39, 72, 73, 203]

GREW, N. (1682). *The Anatomy of Plants. With an Idea of a Philosophical History of Plants*. London. [pp. 32, 37, 72, 143]

GUNTHER, R. T. (1934). *The Greek Herbal of Dioscorides illustrated by a Byzantine* A.D. 512. *Englished by John Goodyer* A.D. 1655. Edited by R. T. G. Oxford. [pp. 27, 123]

HAGERUP, O. (1933). Zur Organogenie und Phylogenie der Koniferen-Zapfen. *Kgl. Danske Vidensk. Selsk. Biol. Medd.* vol. X, 7, 82 pp. [p. 128]

LIST OF BOOKS AND MEMOIRS CITED

HAGERUP, O. (1942). The morphology and biology of the *Corylus*-fruit. *Kgl. Danske Vidensk. Selsk. Biol. Med.* Bd. XVII, no. 6, pp. 1–32. [p. 131]

HALLETT, H. F. (1930). *Aeternitas: A Spinozistic Study.* Oxford. [p. 157]

HAMPTON, F. A. (1925). *The Scent of Flowers and Leaves.* London. [p. 191]

HAYATA, B. (1931). Über das 'Dynamische System' der Pflanzen. *Ber. Dtsch. Bot. Ges.* vol. XLIX, pp. 328–48. [p. 66]

HEALEY, J. ("J. H.".) (1610). *St Augustine, of the Citie of God. Englished by J. H.* Printed by George Eld (no place-name). [p. 2]

HEALEY, J. and TASKER, R. V. G. (1945). *Saint Augustine. The City of God (De Civitate Dei).* Revised by R. V. G. T. from J. H.'s translation, ed. 2 (1620). 2 vols. London: Everyman's Library. [p. 2]

HEDWIG, J. (1781). Vom waren Ursprunge der mänlichen Begattungs-werkzeuge der Pflanzen. *Leipziger Mag. zur Naturk., Math., und Oecon*, Jahrg. 1781, pt. III, pp. 297–319. [p. 31]

HENNELL, T. (1936). *Change in the Farm.* Edition 2. Cambridge. [p. 182]

HETT, W. S. (1935). *Aristotle: On the Soul. Parva naturalia. On Breath.* Loeb Classical Library. London. [pp. 10, 12, 14, 19, 33, 78, 207, 209]

HETT, W. S. (1936). *Aristotle: Minor Works.* Loeb Classical Library. London. [pp. 24, 76]

HIRMER, M. (1932). Die Deutung des weiblichen Blütenzapfens der Coniferen. *Ber. Dtsch. Bot. Ges.* vol. L, pp. (47)–(52). [p. 129]

HOCQUETTE, M. (1946). *Les 'Fantaisies botaniques' de Goethe.* Lille. [p. 64]

HOLLAND, P. (1601). *The Historie of the World. Commonly called, the Naturall Historie of C. Plinius Secundus. Translated into English by P. H.* London. [p. 95]

HORT, A. (1916). *Theophrastus. Enquiry into Plants.* 2 vols. Loeb Classical Library. London. [pp. 12–23, 143, 202]

HUTCHINSON, J. (1935). Evolution of the involucre of bracts in the family Saururaceae. *Proc. Linn. Soc. Lond.* Session 1934–5, pt. III, p. 54. [p. 145]

HUXLEY, J. S. (1932). *Problems of Relative Growth.* London. [p. 168]

HUXLEY, J. S.; *see also* DIVER, C. (1940), *and* SPRAGUE, T. A. (1940).

HUXLEY, T. H. (1853). A Lecture on the identity of structure of plants and animals. (Roy. Inst. of Great Britain). *Quart. Journ. Micr. Sci.* vol. I, pp. 307–11. [p. 196]

HUXLEY, T. H. (1888). The Gentians: Notes and Queries. *Journ. Linn. Soc. Lond. (Bot.)*, vol. XXIV, pp. 101–24. [pp. 64, 65, 160]

JÄGER, G. F. VON (1814). *Ueber die Missbildungen der Gewächse.* Stuttgart. [p. 5]

JESSEN, K. F. W.; *see* MEYER, E. H. F., and JESSEN, K. F. W. (1867).

JOHNSON, F. R. (1937). *Astronomical Thought in Renaissance England: A Study of the English Scientific Writings from 1500 to 1645.* Baltimore. [p. 37]

JOHNSON, M. A. (1934). The origin of the foliar pseudo-bulbils in *Kalanchoë daigremontiana. Bull. Torr. Bot. Club*, vol. LXI, pp. 355–66. [p. 107]

JOHNSON, T.; *see* GERARD, J. (1633).

JOSHI, A. C. (1935). Criticism of Dr Thomas's recent hypothesis on the nature of the angiospermous carpel. *Journ. Bot.* vol. LXXIII, pp. 286–94. [p. 174]

220

LIST OF BOOKS AND MEMOIRS CITED

JUNG, J. (1747). *Opuscula botanico-physica.* Edited by J. S. Albrecht. Coburg. [pp. 33–36, 73, 143]

KANT, I. (1902–38). *Gesammelte Schriften.* Herausgegeben v.d.k. Preuss. Akad. d. Wiss. Berlin. [pp. 73, 203, 208]

KANT, I. (translations); see MEREDITH, J. C. (1928), SMITH, N. KEMP (1933).

KITCHIN, G. W. (1855). *F. Bacon. The Novum Organon.* Translated by G. W. K. Oxford. [pp. 5, 27]

LAND, J. P. N.; see VLOTEN, J. VAN and LAND, J. P. N. (1882, 1883).

LASZLO, P. DE; see WILMOTT, A. J. (1945).

LEAVITT, R. G. (1909). *A vegetative mutant, and the principle of homoeosis in plants. Bot. Gaz.* vol. XLVII, pp. 30–68. [pp. 80, 139]

LILLIE, R. S. (1945). *General Biology and Philosophy of Organism.* Chicago. [pp. 203]

LINNAEUS, C. (LINNÉ, C. VON) (1749). *Amoenitates Academicae.* Vol. I. Lugduni Batavorum. Rudberg, D., *De Peloria,* pp. 280–98. [pp. 185, 190, 197]

LINNAEUS, C. (LINNÉ, C. VON) (1751). *Philosophia Botanica.* Stockholm. [p. 134]

LONES, T. E. (1912). *Aristotle's Researches in Natural Science.* London. [p. 9]

LOVEJOY, A. O. (1936). *The Great Chain of Being.* Harvard University Press. [p. 66]

LOVEJOY, A. O. (1948). *Essays in the History of Ideas.* Baltimore. [p. 174]

LYTE, H. (1578). *A Nievve Herball, or Historie of Plantes . . . nowe first translated out of French into English, by Henry Lyte Esquyer.* London. [p. 4]

MACLEOD, J. (1919, ed. 1, and 1926, ed. 2). *The Quantitative Method in Biology.* London and Manchester. [pp. 155, 171]

McCLURE, F. A. (1934). The inflorescence in *Schizostachyum* Nees. *Journ. Wash. Acad. Sci.* vol. XXIV, no. 12, pp. 541–8. [p. 149]

M'COSH, J. and DICKIE, G. (1856). *Typical Forms and Special Ends in Creation.* Edinburgh. [p. 83]

McDONALD, J. E. (1904). The tuberous moschatel. *Adoxa Moschatellina. Nature Study,* vol. XIII, pp. 189–96. [p. 185]

McINTYRE, W. G. and CHRYSLER, M. A. (1943). The morphological nature of the photosynthetic organs of *Orchyllium Endresii* as indicated by their vascular structure. *Bull. Torr. Bot. Club,* vol. LXX, pp. 252–60. [p. 89]

MAHESHWARI, P. (1937). A critical review of the types of embryo sacs in angiosperms. *New Phytol.* vol. XXXVI, pp. 359–417. [p. 157]

MAHESHWARI, P. (1941). Recent work on the types of embryo-sacs in angiosperms—a critical review. *Journ. Ind. Bot. Soc.* vol. XX, pp. 229–61. [p. 157]

MAHESHWARI, P. (1948). The angiosperm embryo sac. *Bot. Rev.* vol. XIV, pp. 1–56. [p. 157]

MAJUMDAR, G. P. (1942). The organization of the shoot in *Heracleum* in the light of development. *Ann. Bot.,* N.S., vol. VI, pp. 49–81. [p. 125]

MAJUMDAR, G. P. (1947). Growth unit or the phyton in dicotyledons with special reference to *Heracleum. Bull. Bot. Soc., Bengal,* April 1947, pp. 61–6. [p. 166]

LIST OF BOOKS AND MEMOIRS CITED

MALPIGHI, M. (1675, 1679). *Anatome Plantarum.* London. [pp. 37–39, 55, 143]

MALPIGHI, M. (1687). *Opera Omnia.* London. [p. 37]

MALPIGHI, M. (1697). *Opera Posthuma.* London. [p. 37]

MEDAWAR, P. B.; *see* CLARK, W. E. LE GROS and MEDAWAR, P. B. (1945).

MEREDITH, J. C. (1928). *Kant's Critique of Teleological Judgement.* Trans. by J.C.M. Oxford. [pp. 73, 203]

MEYER, A. (1934). *Ideen und Ideale der Biologischen Erkenntnis.* Leipzig. [p. 63]

MEYER, E. H. F. (1854–7). *Geschichte der Botanik.* 4 vols. Königsberg. [p. 24]

MEYER, E. H. F., and JESSEN, K. F. W. (1867). *Alberti Magni...de vegetabilibus libri VII.* Berlin. [p. 24]

MILL, J. S. (1843). *A System of Logic.* 2 vols. London. [p. 160]

MILLER, H. A., and WETMORE, R. H. (1945, 1946). Studies in the developmental Anatomy of *Phlox Drummondii* Hook. *Amer. Journ. Bot.* I. The embryo, vol. XXXII, pp. 588–99; II. The seedling, vol. XXXII, pp. 628–34; III. The apices of the mature plant, vol. XXXIII, pp. 1–10. [pp. 47, 80, 127, 162]

MÖBIUS, M. (1884). Untersuchungen über die Morphologie und anatomie der Monokotylenähnlichen Eryngien. *Prings. Jahrb. wiss. Bot.* vol. XIV, pp. 379–425. [p. 139]

MÖBIUS, M. (1886). Weitere Untersuchungen über Monokotylenähnliche Eryngien. *Prings. Jahrb. wiss. Bot.* vol. XVII, pp. 591–621. [p. 139]

MÖBIUS, M.; *see also* URBAN, I. and MÖBIUS, M. (1884).

MONTAGU, M. F. ASHLEY; *see* ARBER, A. (1946 b).

MONTAIGNE, MICHEL DE (1906 etc.). *Les Essais.* Bordeaux. [p. 6]

MÜLLEROTT, M. (1940). Vergleichende und entwicklungsgeschichtliche Untersuchungen über Zwischenfieder- und Stipellenbildung. *Bot. Archiv,* vol. XL, pp. 258–88. [pp. 35, 123]

MURE, G. R. G. (1926). (In *The Works of Aristotle Translated into English.* Vol. I.) *Analytica Posteriora.* Oxford. [p. 199]

NICHOLSON, R. A. (1921). *Studies in Islamic Mysticism.* Cambridge. [p. 211]

NORMAN, J. M. (1857). *Quelques Observations de Morphologie Végétale.* Programme de l'Université pour le 1er Sem. Christiania. 32 pp. [pp. 99, 141, 153]

NORMAN, J. M. (1858). Quelques observations de morphologie végétale. *Ann. sci. nat.* (*Bot.*), ser. IV, vol. IX, pp. 105–41. [pp. 99, 141, 153]

OAKESHOTT, M. (1933). *Experience and its Modes.* Cambridge. [p. 201]

OGLE, W. (1897). *Aristotle on Youth and Old Age, Life and Death and Respiration.* London. [p. 10]

OGLE, W. (1912). (In *The Works of Aristotle Translated into English.* Vol. V.) *De partibus animalium.* Oxford. [pp. 8, 10, 11, 202]

OKEN, L. (1809–11). *Lehrbuch der Naturphilosophie.* 3 vols. Jena. [pp. 83, 87, 158]

OKEN, L. (1847). *Elements of Physiophilosophy.* Trans. by A. TULK. Ray Society. London. [pp. 83, 87]

OWEN, R. (1894). *The Life of Richard Owen.* 2 vols. London. [p. 67]

LIST OF BOOKS AND MEMOIRS CITED

PECK, A. L. (1937). *Aristotle. Parts of Animals.* Loeb Classical Library. London. [pp. 8, 11, 12, 196]

PENZIG, O. (1890, 1894). *Pflanzen-Teratologie.* 2 vols. Genoa. [p. 191]

PENZIG, O. (1921, 1922). *Pflanzen-Teratologie.* Edition 2. 3 vols. Berlin. [pp. 105, 129]

PHILIPSON, W. R. (1946). Studies in the development of the inflorescence. I. The capitulum of *Bellis perennis* L. *Ann. Bot.,* N.S., vol. x, pp. 257–70. [pp. 47, 173]

PHILIPSON, W. R. (1947 a). Studies in the development of the inflorescence. II. The capitula of *Succisa pratensis* Moench. and *Dipsacus fullonum* L. *Ann. Bot.,* N.S., vol. xi, pp. 285–97. [p. 127]

PHILIPSON, W. R. (1947 b). *Some observations on the apical meristems of leafy and flowering shoots. Journ. Linn. Soc. (Bot.),* vol. LIII, pp. 187–93. [p. 47]

PHILIPSON, W. R. (1949). The ontogeny of the shoot apex in Dicotyledons. *Biol. Rev.* vol. XXIV, pp. 21–50. [p. 135]

PILGER, R. (1926). Gymnospermae. In *Die natürlichen Pflanzenfamilien* (Engler, A. and Prantl, K.), ed. 2, vol. XIII, Leipzig. [p. 128]

PLATO (translation); *see* CORNFORD, F. M. (1937).

PLINY; *see* HOLLAND, P. (1601) and SILLIG, J. (1851–8).

PLOTINUS; *see* BRÉHIER, E. (1927).

POIRET, J. L. M.; *see* TURPIN, P. J. F. (1820).

POLLOCK, F. (1899). *Spinoza: his life and philosophy.* Edition 2. London. [p. 76]

POPHAM, R. A. (1947). Developmental anatomy of seedling of *Jatropha cordata. Ohio Journ. Sci.* vol. XLVI, pp. 1–20. [p. 30]

POTONIÉ, N. (1899). *Lehrbuch der Pflanzenpalaeontologie.* Berlin. [p. 74]

PRIESTLEY, J. H., SCOTT, L. I. and GILLETT, E. C. (1935). The development of the shoot in *Alstroemeria* and the unit of shoot growth in monocotyledons. *Ann. Bot.* vol. XLIX, pp. 161–79. [p. 166]

PRIESTLEY, J. H.; *see also* UNDERWOOD, D. and SCOTT, L. I. (1935).

PRINGSHEIM, E. G. (1932). *Julius Sachs. Der Begründer der neueren Pflanzenphysiologie. 1832–1897.* Jena. [pp. 66, 155, 160, 205, 210]

PURVIS, O. N. and GREGORY, F. G. (1937). Studies in vernalisation of cereals. I. A comparative study of vernalisation of winter rye by low temperature and by short days. *Ann. Bot.,* N.S., vol. I, pp. 569–91. [p. 55]

RAISTRICK, H. (1932). Biochemistry of the lower fungi. *Ergeb. Enzymforsch.* vol. I, pp. 345–63. [p. 195]

RAND, E. K.; *see* STEWART, H. F. and RAND, E. K. (1918).

RAUH, W. (1937). Die Bildung von Hypocotyl- und Wurzelsprossen. *Nova Acta Leopoldina,* N.S., vol. IV, pp. 395–553. [pp. 30, 132]

RENAN, E. (1852). *Averroès et l'Averroïsme.* Paris. [p. 28]

RENNIE, J. (1849). *A Familiar Introduction to Botany.* A new edition revised. London. [pp. 12, 30]

RICHARDS, O. W.; *see* ROBSON, G. C. and RICHARDS, O. W. (1936).

ROBERTS, M. (1926). *Malignancy and Evolution: A Biological Inquiry into the Nature and Causes of Cancer.* London. [p. 11]

LIST OF BOOKS AND MEMOIRS CITED

ROBERTSON (ARBER), A. (1906). Some points in the morphology of *Phyllo-cladus alpinus*, Hook. *Ann. Bot.* vol. xx, pp. 259–65. [p. 127]

ROBINET, J. B. R. (1768a). *Considérations philosophiques de la gradation naturelle des formes de l'être.* Paris. [pp. 66, 67]

ROBINET, J. B. R. (1768b). *Vue philosophique de la gradation naturelle des formes de l'Être.* Amsterdam.¹ [pp. 66, 67]

ROBSON, G. C. and RICHARDS, O. W. (1936). *The Variation of Animals in Nature.* London. [p. 63]

RUDBERG, D.; see LINNAEUS, C. (1749).

RUSSELL, E. S. (1916). *Form and Function.* London. [p. 3]

RUSSELL, E. S. (1930). *The Interpretation of Development and Heredity: A Study in Biological Method.* Oxford. [p. 196]

RUSSELL, E. S. (1932). Conation and perception in animal learning. *Biol. Rev.* vol. VII, pp. 149–79. [p. 211]

RUSSELL, E. S. (1933). The Limitations of Analysis in Biology. (Meeting of the Aristotelian Society, Feb. 27, 1933.) *Proc. Arist. Soc.*, N.S., vol. XXXIII, 1932–3, pp. 147–58. [p. 158]

RUSSELL, E. S. (1934). The Study of Behaviour. Pres. Add. to Sect. D, Zoology, *Rep. Brit. Ass. Adv. Sci.*, Aberdeen, pp. 83–98. [p. 3]

RUSSELL, E. S. (1945). *The Directiveness of Organic Activities.* Cambridge. [pp. 78, 205]

SACHS, J. VON² (1868). *Lehrbuch der Botanik.* Leipzig. [pp. 71, 83, 129]

SACHS, J. VON (1870). *Lehrbuch der Botanik.* Edition 2. Leipzig. [p. 71]

SACHS, J. VON (1875). *Text-Book of Botany.* Trans. (from the 3rd and 4th editions) by A. W. Bennett and W. T. Thiselton Dyer. Oxford. [pp. 71, 83, 142]

SACHS, J. VON (1890). *History of Botany* (1530–1860). Trans. by H. E. F. Garnsey and I. B. Balfour. Oxford. [p. 30]

SACHS, J. VON (1893). Physiologische Notizen. VI. Ueber einige Beziehungen der specifischen Grösse der Pflanzen zu ihrer Organisation. VII. Ueber Wachsthumsperioden und Bildungsreize. *Flora*, vol. LXXVII, pp. 49–81, 217–53. [pp. 190, 191]

SAHNI, B. (1920). On the structure and affinities of *Acmopyle Pancheri*, Pilger. *Phil. Trans. Roy. Soc.*, Lond., B, vol. CCX, pp. 253–310. [pp. 91, 127, 128]

SAINT-HILAIRE; see GEOFFROY SAINT-HILAIRE, I. (1832–6).

SALAMAN, R. N. (1946). The Early European Potato: its character and place of origin. *Journ. Linn. Soc. Lond.* (*Bot.*), vol. LIII, pp. 1–27. [p. 23]

SAMASSA, P. (1896). Translation of Wolff, C. F., *Theoria generationis*, 1759. Ostwald's *Klassiker der Exakten Wissenschaften*, nos. 84, 85. Leipzig. [p. 48]

¹ Except for the variant in title, Robinet, J. B. R. (1768a) and (1768b) appear to be identical.

² The full citation should be Sachs, F. G. J. von, but as his botanical work was done under the name Julius alone, it seems better to refer to him as Sachs, J. von.

LIST OF BOOKS AND MEMOIRS CITED

SAVORY, T. H. (1936). *Mechanistic Biology and Animal Behaviour.* London. [p. 201]

SCHAEPPI, H. (1935). Untersuchungen über die Blattentwicklung bei *Ceratophyllum, Cabomba* und *Limnophila. Planta,* vol. XXIV, pp. 755–69. [p. 116]

SCHLECHTENDAL, D. F. L. VON; see CHAMISSO, A. DE and SCHLECHTENDAL, D. F. L. VON (1826).

SCHLEIDEN, M. J. (1848). *Die Pflanze und ihr Leben.* Leipzig. [p. 62]

SCHONLAND, S. (1921). A new genus of Crassulaceae. *Ann. Bolus Herb.* vol. III, pt. II, pp. 67–9. [p. 148]

SCHOUTE, J. C. (1935). On corolla aestivation and phyllotaxis of floral phyllomes. *Ver. Akad. Wet., Amst. Afd. Natuurkunde* (Sect. II), pt. XXXIV, no. 4, 77 pp. [p. 51]

SCHOUTE, J. C. (1937). On the phyllotaxis of the *Ulmus* seedling. *Rec. Trav. Bot. Néerl.* vol. XXXIV, pp. 615–22. [p. 89]

SCHÜEPP, O. (1938). Ueber periodische Formbildung bei Pflanzen. *Biol. Rev.* vol. XIII, pp. 59–92. [pp. 79, 142]

SCHULTZ-SCHULTZENSTEIN, K. H. VON (1861). Die Bedeutung der Verzweigung im Pflanzenreich. *Flora,* N.R. Jahrg., XIX (G.R. XLIV), pp. 273–88, 297–302. [pp. 74, 111]

SCOTT, L. I.; see PRIESTLEY, J. H., SCOTT, L. I. and GILLETT, E. C. (1935); UNDERWOOD, D. and SCOTT, L. I. (1935).

SENN, G. (1930). Hat Aristoteles eine selbständige Schrift über Pflanzen verfasst? *Philologus,* Leipzig, vol. LXXXV (N.F. XXXIX), pp. 113–40. [p. 9]

SENN, G. (1933). Die Entwicklung der biologischen Forschungsmethode in der Antike und ihre grundsätzliche Förderung durch Theophrast von Eresos. *Veröff. Schweiz. Ges. Geschichte Med. Naturwiss.* vol. VIII, 262 pp. [pp. 11, 13]

SHARROCK, R. (1660). *The History of the Propagation and Improvement of Vegetables By the Concurrence of Art and Nature.* Oxford. [pp. 6, 164]

SHERRINGTON, C. (1940). *Man on his Nature.* The Gifford Lectures, Edinburgh, 1937–8. Cambridge. [p. 3]

SIFTON, H. B. (1944). Developmental morphology of vascular plants. *New Phytol.,* vol. XLIII, pp. 87–129. [p. 162]

SILLIG, J. (1851–8). *C. Plinii Secundi Naturalis Historiae Libri XXXVII.* Ed. J. S. Hamburg and Gotha. [p. 95]

SINNOTT, E. W. and BAILEY, I. W. (1914). Investigations on the phylogeny of the angiosperms. 3. Nodal anatomy and the morphology of stipules. *Amer. Journ. Bot.* vol. I, pp. 441–53. [p. 141]

SKUTCH, A. F. (1930). On the development and morphology of the leaf of the banana (*Musa sapientum* L.). *Amer. Journ. Bot.* vol. XVII, pp. 252–71. [p. 85]

SMITH, MARGARET (1944). *Al-Ghazālī the Mystic. A Study of the Life and Personality of Abū Hāmid Muahmmad al-Tūsī al-Ghāzāli, together with an account of his Mystical Teaching and an estimate of his place in the History of Islamic Mysticism.* London. [p. 202]

SMITH, N. KEMP (1933). *Immanuel Kant's Critique of Pure Reason,* Trans. by N. K. S. London. [p. 208]

LIST OF BOOKS AND MEMOIRS CITED

Snow, R. (1929). The young leaf as the inhibiting organ. *New. Phytol.* vol. xxviii, pp. 345–58. [p. 97]

Soret, F.; *see* Goethe, J. W. von (1831).

Spencer, H. (1867). *The Principles of Biology.* Vol. ii. London. [pp. 166, 168]

Spieghel (Spigelius), A. (1606). *Isagoges in rem herbariam Libri Duo.* Padua. [pp. 40, 72]

[Spinoza, B. de] "B. D. S" (1667). *Opera posthuma.* (No place-name.) [pp. 77, 197, 208]

Spinoza, B. de (works and translations); *see* White, W. Hale, and Stirling, A. H. (1899), (1930); Wolf, A. (1910), (1928); Vloten, J. van and Land, J. P. N. (1882, 1883).

Sprague, T. A. (1927). The morphology and taxonomic position of the Adoxaceae. *Journ. Linn. Soc. Lond. (Bot.),* vol. xlvii, pp. 471–87. [p. 185]

Sprague, T. A. (1933a). Plant morphology in Albertus Magnus. *Kew Bull.* no. 9, pp. 431–40. [pp. 24, 27]

Sprague, T. A. (1933b). Botanical terms in Albertus Magnus. *Kew Bull.* no. 9, pp. 440–59. [pp. 24, 73]

Sprague, T. A. (1940). Taxonomic botany, with special reference to the angiosperms. In *The New Systematics,* edited by J. S. Huxley, Oxford, pp. 435–54. [p. 210]

Sprotte, K. (1940). Untersuchungen über Wachstum und Nervatur der Fruchtblätter. *Bot. Arch.* vol. xl, pp. 463–506. [p. 58]

Stace, W. T. (1920). *A Critical History of Greek Philosophy.* London. [p. 199]

Stace, W. T. (1924). *The Philosophy of Hegel. A Systematic Exposition.* London. [pp. 71, 87, 158, 161, 205]

Stefanoff, B. (1932). Ueber das morphologische Wesen der Phyllokladien bei *Asparagus* L. *Bull. Soc. Bot. Bulg.* vol. v, pp. 63–77. [p. 97]

Stewart, H. F. and Rand, E. K. (1918). *Boethius. The Theological Tractates,* translated by H. F. S. and E. K. R., and *The Consolation of Philosophy,* translated by "I. T." (1609), revised by H. F. S. Loeb Classical Library. London. [pp. 12, 77]

Stewart, L. B.; *see* Graham, R. J. D. and Stewart, L. B. (1929).

Stiles, W.; *see* Brooks, F. T. and Stiles, W. (1910).

Stirling, A. H.; *see* White, W. Hale and Stirling, A. H. (1899, 1930).

Stocks, J. L. (1938). *Time, Cause and Eternity.* London. [p. 201]

Stocks, J. L. (1939). *Reason and Intuition.* Oxford. [p. 211]

Strasburger, E. (1872). *Die Coniferen und die Gnetaceen.* Leipzig. [p. 129]

Strömberg, R. (1937). Theophrastea. *Vetensk.-Samh. Handl., Göteborg.* Föl. v., ser. A, vol. vi, no. 4, 234 pp. [pp. 11, 13, 15, 16, 17, 210]

Swift, J. (1727). *Miscellanies in Prose and Verse.* Vol. ii. London. [p. 71]

Takhtajan (Tacktajan), A. (1943). Correlations of ontogenesis and phylogenesis in higher plants. *Trans. Molotov State Univ. of Erevan,* vol. xxii, pp. 71–176. (Russian, with English summary, pp. 168–73.) [p. 108]

LIST OF BOOKS AND MEMOIRS CITED

TANSLEY, A. G. (1908). Lectures on the evolution of the filicinean vascular system. *New Phytol. Reprints,* no. 2, Cambridge. [pp. 74, 91]

TANSLEY, A. G. (1948). The nature and range of variation in the floral symmetry of *Potentilla erecta* (L.) Hampe. *New Phytol.* vol. XLVII, pp. 95–110. [p. 164]

TASKER, R. V. G.; *see* HEALEY, J. and TASKER, R. V. G. (1945).

TAYLOR, A. E. (1929). *Plato: the Man and his Work.* Edition 3. London. [p. 211]

THEOPHRASTUS (translations); *see* DENGLER, R. E. (1927) and HORT, A. (1916).

THODAY, D. (1939). The interpretation of plant structure. Pres. Add. to Sect. K, Bot., *Rep. Brit. Ass. Adv. Sci.,* Dundee. New Quarterly Series, I, 1939–40, pt. I, 1939, pp. 84–104. [p. 134]

THOMAS, H. HAMSHAW (1932). The old morphology and the new. *Proc. Linn. Soc. Lond.* Session 145 (1932–3), pp. 17–32. [p. 76]

THOMAS, H. HAMSHAW (1947). *The History of Plant Form* (Brit. Ass. Adv. Sci.; Pres. Add. Sect. K, Botany, Dundee). *The Advancement of Science,* vol. IV, no. 15, pp. 243–54. [p. 66]

THOMPSON, D'ARCY W. (1910). (In *The Works of Aristotle Translated into English.* Vol. IV.) *Historia animalium.* Oxford. [pp. 11, 61]

THOMPSON, D'ARCY W. (1913). *On Aristotle as a Biologist.* Herbert Spencer Lecture. Oxford. [pp. 9, 14]

THOMPSON, D'ARCY W. (1917, 1942). *On Growth and Form.* Edition 1, 1917; edition 2, 1942. Cambridge. [pp. 9, 168, 196, 201]

THOMSON, R. B. (1940). The structure of the cone in the Coniferae. *Bot. Rev.* vol. VI, pp. 73–84. [p. 129]

TRÉCUL, A. (1853). Mémoire sur la formation des feuilles. *Ann. sci. nat. (Bot.),* ser. III. vol. XX, pp. 235–314. [p. 79]

TREDENNICK, H. (1933). *Aristotle. The Metaphysics. I–IX.* Trans. by H. T. Loeb Classical Library. London. [pp. 199, 204, 207]

TREDENNICK, H. (1935). *Aristotle. The Metaphysics. X–XIV.* Trans. by H. T. Loeb Classical Library. London. [pp. 204, 207]

TROLL, W. (1926). *Goethes Morphologische Schriften.* Jena. [pp. 4, 40, 41, 59, 61, 93, 132, 136, 206, 209, 210]

TROLL, W. (1928a). *Organisation und Gestalt im Bereich der Blüte.* (Monogr. wiss. Bot. 1.) Berlin. [pp. 144, 147, 155, 209]

TROLL, W. (1928b). Zur Auffassung des parakarpen Gynaeceums und des coenokarpen Gynaeceums überhaupt. *Planta,* vol. VI, pp. 255–76. [p. 58]

TROLL, W. (1931–3). Beiträge zur Morphologie des Gynaeceums. I. Ueber das Gynaeceum der Hydrocharitaceen. *Planta,* vol. XIV, 1931, pp. 1–18; II. Ueber das Gynaeceum von *Limnocharis* Humb. et Bonpl. Ibid. vol. XVII, 1932, pp. 453–60; III, Ueber das Gynaeceum von *Nigella,* und einiger anderer Helleboreen. Ibid. vol. XXI, 1933, pp. 266–91; IV, Ueber das Gynaeceum der Nymphaeaceen. Ibid. vol. XXI, 1933, pp. 447–85. [p. 58]

TROLL, W. (1932). Morphologie der schildförmigen Blätter. *Planta,* vol. XVII, pp. 153–314. [pp. 58, 79, 109]

TROLL, W. (1934*a*). Ueber den Bau der Rhachis und seinem Einfluss auf die Spreitenbildung von Fiederblättern. *Planta*, vol. XXII, pp. 80–108. [p. 116]

TROLL, W. (1934*b*). Ueber Bau und Nervatur der Karpelle von *Ranunculus*. *Ber. Dtsch. Bot. Ges.* vol. LIII, pp. 214–20. [p. 58]

TROLL, W. (1935, etc.).[1] *Vergleichende Morphologie der höheren Pflanzen*. Berlin (in course of publication). [pp. 51, 58, 68, 70, 71, 82, 85, 111, 159]

TULK, A.; *see* OKEN, L. (1847).

TURPIN, P. J. F. (1820). Iconographie Végétale. In Poiret, J. L. M., *Leçons de Flore*. Vol. III. Paris. [pp. 70, 210]

TURPIN, P. J. F. (1837). *Esquisse d'organographie végétale...pour servir à prouver...la métamorphose des plantes de Goethe*. Paris and Geneva. [pp. 41, 62]

UITTIEN, H. (1928*a*). Ueber den Zusammenhang zwischen Blattnervatur und Sprossverzweigung. *Rec. Trav. bot. néerland.* vol. XXV, pp. 390–483. [pp. 83, 102, 119, 184]

UITTIEN, H. (1928*b*). Ueber eine abweichende Form von *Anthriscus sylvestris* Hoffm. *Rec. Trav. bot. néerland.* vol. XXV*a*, pp. 445–51. [p. 119]

UNDERWOOD, D. and SCOTT, L. I., in PRIESTLEY, J. H. and SCOTT, L. I. (1935). The bud scale. *The Naturalist*, pp. 219–28. [p. 178]

UNDERWOOD, E. ASHWORTH; *see* ARBER, A. (forthcoming).

URBAN, I. and MÖBIUS, M. (1884). Ueber *Schlechtendalia luzulifolia*, Less., eine Monocotylenähnliche Composite, und *Eryngium eriophorum*, Cham., eine grasblättrige Umbellifere. *Ber. Dtsch. Bot. Ges.*, vol. II, pp. 100–7. [p. 139]

VELENOVSKÝ, J. (1905–13). *Vergleichende Morphologie der Pflanzen*. 4 vols. Prague. [pp. 83, 99, 182]

VIALLETON, L. (1930). *L'origine des êtres vivants. L'illusion transformiste*. Paris. [p. 66]

VLOTEN, J. VAN and LAND, J. P. N. (1882, 1883). *Benedicti de Spinoza Opera*. 2 vols. Hague. [p. 77]

WALTON, J. (1929). On the structure of a palaeozoic cone-scale and the evidence it furnishes of the primitive nature of the double cone-scale in the conifers. *Mem. Manchester Lit. Phil. Soc.* vol. LXXIII, 1929, for 1928–9, pp. 1–6. [p. 128]

WARDLAW, C. W. (1924–8). Size in relation to internal morphology. *Trans. Roy. Soc. Edinb.* vol. LIII, 1925, pt. III, 1924, pp. 503–32; vol. LIV, 1926, pt. II, 1925, pp. 281–308; vol. LVI, 1931, pt. I, 1928, pp. 19–55. [p. 190]

WARMING, E. (1872). Forgreningsforhold hos Fanerogamerne, betragtede med saerligt Hensyn til Kløvning af Vaekstpunktet. *Vidensk. Selsk. 5.* Række, Naturvidens. og Math. Afd. Kjøbenhavn. Afd. 10B, 1., 173 + L pp. (French summary: *Recherches sur la ramification des Phanérogames principalement au point de vue de la partition du point végétatif.*) [p. 125]

---

[1] The date on the title-page of the first volume is 1937, but the first part is dated 1935.

LIST OF BOOKS AND MEMOIRS CITED

WEISS, F. E. (1932). Some unusual female catkins of *Corylus. Proc. Linn. Soc. Lond.*, 1931–2, pp. 107–9. [p. 178]

WETMORE, R. H. (1943). Leaf-stem relationships in the vascular plants. *Torreya*, vol. XLIII, pp. 16–28. [p. 166]

WETMORE, R. H.; *see also* MILLER H. A. and WETMORE, R. H. (1945, 1946).

WHELDALE, M. (1916). *The Anthocyanin Pigments of Plants.* Cambridge. [p. 184]

WHEWELL, W. (1840). *The Philosophy of the Inductive Sciences.* 2 vols. London. [p. 67]

WHITE, W. HALE, and STIRLING, A. H. (1899). *Spinoza. Tractatus de Intellectus Emendatione.* Trans. by W. H. W., revised by A. H. S. London. [pp. 1, 77]

WHITE, W. HALE, and STIRLING, A. H. (1930). *Benedict de Spinoza, Ethic.* Trans. by W. H. W. and A. H. S. Edition 4. Oxford. [p. 77]

WHITTAKER, T. (1918). *The Neo-Platonists.* Edition 2. Cambridge. [pp. 2, 209]

WILLIS, J. C. (1902). Studies in the morphology and ecology of the Podostemaceae of Ceylon and India. *Ann. Roy. Bot. Gdns Peradeniya*, vol. I, pp. 267–465. [pp. 133, 135]

WILMOTT, A. J. (1945). Remarks in a discussion on Laszlo, P. de, Colour photography applied to biology. *Proc. Linn. Soc. Lond.* Session 157, 1944–5, pp. 61–2. [p. 211]

WILSON, C. L. (1942). The telome theory and the origin of the stamen. *Amer. Journ. Bot.* vol. XXIX, pp. 759–64. [p. 76]

WOLF, A. (1910). *Spinoza's Short Treatise on God, Man, and his Well-being.* Trans. and ed. with a *Life of Spinoza*, by A. W. London. [pp. 77, 78]

WOLF, A. (1928). *The Correspondence of Spinoza.* Trans. and ed. by A. W. London. [p. 197]

WOLFF, C. F. (1768). De formatione intestinorum. *Novi Commentarii Acad. Sci. Imperialis Petropolitanae*, vol. XII, pp. 403–507. [p. 40]

WOLFF, C. F.; *see also* SAMASSA, P. (1896).

WOLFF, H. (1913). Umbelliferae-Saniculoideae. In Engler, A., *Das Pflanzenreich*, vol. IV, p. 228. [p. 139]

WOLFSON, H. A. (1947). *Philo.* 2 vols. Cambridge, Mass. [p. 65]

WOODGER, J. H. (1929). *Biological Principles: A Critical Study.* London. [p. 176]

WORSDELL, W. C. (1900). The structure of the female 'flower' in Coniferae. *Ann. Bot.* vol. XIV, pp. 39–82. [p. 128]

WYDLER, H. (1844). Morphologische Mittheilungen. 5. Verzweigung der Solaneen. *Bot. Zeit. Jahrg.* II, pp. 689–94, 705–8. [p. 95]

WYDLER, H. (1857). Ueber die Blatt- und Blüthenstellung von *Solanum nigrum* und den Verwandten Arten. *Flora*, N.R., Jahrg. XV, G.R., XL, pp. 225–33. [p. 95]

ZIMMERMANN, W. (1930). *Die Phylogenie der Pflanzen: Ein Überblick über Tatsachen und Probleme.* Jena. [pp. 75, 155, 191]

INDEX

(For authors' names not included in this index see *List of Books and Memoirs Cited*, pp. 212–229, where page references to citations in the text will be found. As a rule, botanical family names are indexed only when several of their genera are named in the text.)

INDEX

Bindweed, see *Calystegia sepium* R.Br.

Blackberry, see *Rubus fruticosus* L.

Blite, see *Amaranthus Blitum* L.

Blue and yellow coloration, distribution in flowers and inflorescences, 147, 191

Bock, J. (Tragus, H.), compared with Malpighi and Grew, 37

Boethius, on self-maintenance in the plant world, 76, 77; on the root corresponding to the head, 12

Borage, Bacon, F., on, 27

Boraginaceae; *Borago*, 27; *Myosotis*, 147, 191

Bower, F. O., on 'Size-Factor', 190

Bracteoles, equivalent to prophylls, 48

Brassica oleracea L., ascidial peltate lamina, 109, 110 (Fig. 15, B1–B3); enations from lamina, 82; lamina-wings from midrib, 107, 108; shoot-like outgrowth from midrib, 110 (Fig. 15, B 4); tuft of laminae borne by leaf, 106 (Fig. 14, B)

Brassica sp., 115 (Fig. 18, A)

Braun, A., short-shoot theory of the ovuliferous scale in conifers, 128–130

Brieger, F. G., on chemical factors in the differentiation of the flower, 191 *n.*

Brown, R., on connation, 171; on reticulate connexion of taxonomic groups, 66; on the relation of the leaf to the parts of the flower, 68; connexion with Corréa da Serra, 60

Browne, Sir T., on the rose calyx, 50

Bud, axillary, a new interpretation of bud morphology, 125–31 (summarised, 126); anatomical connexion with parent shoot, 88 (Fig. 7, B), 91, 92, 126, 127; inhibiting action of leaves on axillary and terminal bud, 97

'Buglosse', Bacon, F., on, 27

Buitenzorg Garden, 149 *n.*

Bulbous plants, Cesalpino on gynaecea of, 29

Bundles, collateral, replaced by radial strands or steles, 104, 105, 106 (Fig. 14, A)

'Bundles of insertion' for axillary bud, 88 (Fig. 7, B), 91, 92, 126, 127

Bupleurum rotundifolium L., pseudanthia 149, 151, 152 (Fig. 29, A)

Butchers'-broom, see *Ruscus aculeatus* L.

Buttercup, see *Ranunculus*; Water-buttercup, see *R. heterophyllus* Fries

Cabbage, see *Brassica oleracea* L.

Caesalpinia japonica Sieb. et Zucc., 89, 90 (Fig. 8, A)

Caesalpinia sepiaria Roxb., see *C. japonica* Sieb. et Zucc.

Calendula officinalis L., 153, 154 (Fig. 30, C)

Calystegia sepium R.Br., leaf with second apex, 101 (Fig. 12, C), 104; Theophrastus on the corolla, 20

Cambridge Platonists, 202

Camellia japonica L., transitions from stamens to petals, 54 (Fig. 4, C), 55

Campanula, Jung on heterophylly, 35; petaloid calyx, 51, 53 (Fig. 3, B)

Campanula rotundifolia L., palmately and pinnately veined leaves, 101 (Fig. 12, A), 102

Candolle, A. Casimir P. de, on the nature of the leaf, 74 *et passim*; relationship to Candolle, A. P. de, 74

Candolle, A. P. de, his ideas in relation to those of Goethe and Grew, 46; on abnormalities, 5; on corolla-androecium system, 55; on family archetypes, 63; on Goethe, 41; on parallelism of vegetative and floral members, 76; on sectorial analysis of shoot, 166; on symmetry, 174, 176, 177; on the description of marginal 'cutting' in leaves, 83, 84; on the flower, 46, 47; on the nœud vital, 30; on units of plant structure, 70

Canna, Goethe on, 44

Cannabis sativa L., Cesalpino on sterile and fertile flowers, 29

Canterbury-bells, see *Campanula*

Caper, see *Capparis spinosa* L.

Capparidaceae, *Capparis*, 56, 57 (Fig. 5, D); *Gynandropsis*, 48, 49 (Fig. 1, D), 88 (Fig. 7, B), 92, 175 (Fig. 37, D); *Polanisia*, 51, 53 (Fig. 3, A)

Capparis spinosa L., stamen anatomy, 56, 57 (Fig. 5, D)

Caput radicis of Cesalpino, 29, 30

Carpels, corresponding to peltate leaves 58, 111; interpretation of, 44–6, 56, 58, 129–31; non-vascular, 58; terminal, 124; theory of Delpino, 129–31

Carrot, see *Daucus Carota* L.

Carum Petroselinum Benth. et Hook., Jung on branch-like leaves, 73

INDEX

Compound-pinnate leaves, comparison with shoots, 114, 116, 118; interpretation, 119–23
Conatus of Spinoza, 77, 207
Conditions *versus* causes, 196–8
'Congenital fusion', 171
Conifers, female reproductive shoot, 127–31
Connation, 171
Convolvulus, see *Calystegia sepium* R.Br.
Coppicing, effect on *Corylus Avellana* L., 177, 178
Cor plantarum of Cesalpino, 29, 30
Corchorus olitorius L., hair-like lobes and stipules, 140 (Fig. 26, D), 141
Coriandrum sativum L., pseudanthia, 151, 152 (Fig. 29, B)
Coriaria myrtifolia L., leaf-like branch system and stipular hairs, 89, 90 (Fig. 8, B), 142
Corn, Theophrastus on the ear, 143
'Coronary' herbs, Cesalpino on, 31
Corréa da Serra, J. F., life, 59, 60; on a plan underlying each plant group, 59–61; on parallelism, 160; on symmetry, 174; on vegetation of New Holland, 60
Corylus Avellana L., effect of coppicing, 177, 178; Theophrastus on the male catkin, 20, 21
Cotyledons, and prophylls, 70, 71; and stipules, 80–2; Goethe on, 43; root-shoot members, 30
Cow-parsley, see *Anthriscus sylvestris* Hoffm.
Crambe maritima L., inflorescences replacing flowers, 153, 154 (Fig. 30, A); narrowing of fruit base due to pressure, 197
Crane's-bill, Long-stalked, see *Geranium columbinum* L.
Crassulaceae, *Pagella*, 148
Crataegus Oxyacantha L., asymmetrical stipules, 52 (Fig. 2, A 3), 82; stipulate and exstipulate leaves, 51, 52 (Fig. 2, A 1, A 2)
Crocus sativus L., 153; *C.* sp., 203
Cross, G. L., on comparison of scale-leaves and foliage-leaves, 86
Cruciferae, 'bractless' raceme, 99, 127, 140 (Fig. 26, E): effect of pressure in flower development, 173, 175 (Fig. 37, C), 197; range of form in gynaeceum, 168, 170; *Brassica*, 80, 105, 106

(Fig. 14, B), 107–110 (Fig. 15, B), 115 (Fig. 18, A); *Brassica* (turnip), 31; *Cheiranthus*, 175 (Fig. 37, C); *Crambe*, 153, 154 (Fig. 30, A), 197; *Nasturtium*, 99, 140 (Fig. 26, E), 178, 179 (Fig. 39, B), 180; *Succowia*, 169 (Fig. 35, H), 170
Cryptocoryne, pseudanthia, 145
Cryptogams, vascular, the plant body in, 74–6, 91, 116, 139
Cryptomeria japonica (L.-f.) Don, Hagerup on, 128
Cudweed, see *Filago germanica* L.
Cudworth, R., on mechanism and teleology, 202; on "Plastick Nature", 202 *n.*; on the formal cause, 205
Cupressus sempervirens L., Theophrastus on fruit and seed, 22
Cyclamen, Cesalpino on, 31
Cydonia oblonga Mill., Grew on leaf-like sepals, 39
Cymose branching, 94
Cypella Herberti Herb., foliated lamina, 85
Cyperus longus L., Theophrastus on, 19
Cypress, see *Cupressus sempervirens* L.

Daisy, see *Bellis perennis* L.
Daisy, Oxeye, see *Chrysanthemum Leucanthemum* L.
D'Aléchamps (Dalechamps), J., emendation of Theophrastus, 15
Danae, interpretation of sterile phylloclade, 97
Dante, 8, 39, 208
Darlingtonia californica Torr., 'fishtail' appendage, 195
Darstellung, 209
Darwin, C., on analogous variation, 160
Darwinian theory, see Pre- and Post-Darwinian morphology
Date palm, see *Phoenix dactylifera* L.
Daucus Carota L., luxuriant umbel, 149, 150 (Fig. 28); pseudanthia, 151, 152 (Fig. 29, C); terminal umbellule, 182, 183 (Fig. 41), 184, 185, 190
De causis plantarum (Theophrastus) 13, 20, 22, 23
De historia plantarum (Theophrastus), 13–23
Delpino, F., foreshadowing *Gestalt* concept, 155; interpretation of ovuliferous scale in conifers, and of the carpel in angiosperms, 129–31

234

INDEX

Jäger, G. F. von, on abnormalities, 5
Jasminum humile L., range of leaf form, and correspondence of shoot and leaf, Frontispiece, A, and 139
Johnson, T., on Aristotle, 9
Joshi, A. C., on the effect of pressure on ovule development, 174
Judas tree, see *Cercis Siliquastrum* L.
Juncaginaceae, *Lilaea*, 95
Jung, J., *De Plantis Doxoscopiae*, 33; *Isagoge Phytoscopica*, 33; on botanical terminology, 34, 35; on branching, 33; on branch-like leaves, 73; on composite flowers and their 'doubling', 35, 36; on heterophylly, 35; on leaf form, 34, 35; on pappus of composites, 36; on physical and chemical causes, 27; on radial stem and dorsiventral leaf, 34; on subsidiary leaflets in *Spiraea Ulmaria* L., etc., 35; on the 'flower', 35; on the *fundus*, 33; on the root-crown, 33, 143; representing Aristotelian botany, 24

Kalanchoë daigremontiana R. Hamet et Perrier de la Bâthie, leaf-budding, 107, 108 (Fig. 14, C)
Kant, I., on necessity and free will, 208; on parallelism of leaf and shoot, 73, 74; on teleology and mechanism, 203
καρδία, 12

Labiatae, *Ballota*, 163 (Fig. 32, B), 164; *Marrubium*, 28; *Nepeta*, 192 (Fig. 45, D), 193; *Ocimum*, 16 n.; *Salvia*, 169 (Fig. 35, F), 170; *Scutellaria*, 169 (Fig. 35, G), 170; *Thymus*, 22, 192 (Fig. 45, C), 193
Lapsana communis L., alternate pinnae and marginal relation of pinna to rachis, 80, 114, 115 (Fig. 18, C)
Lateral structures taking up parental role, 93 *et seq.*; summarised, 105
Lathyrus Aphaca L., stipules dominating leaf, 99, 100 (Fig. 11)
Laurel, Alexandrian, see *Ruscus Hypophyllum* L.
Leaf, apex dominated by laterals, 97, 98; budding, 105 *et seq.*; Candolle, A. P. de, on leaf, 46, 83, 84; Candolle, Casimir de, on leaf, 74 *et passim*; Cesalpino on leaf, 28, 29; classical view of function of leaf, 17, 25, 28; compound-peltate and compound-

pinnate, comparison with shoot, 111, 114, 116, 118; connexion of generations in leaf, 142, 143; cuttings, 107; dorsiventrality, 18, 34, 87 *et seq.*, 200; Goethe on leaf, 42 *et passim*; Grew on leaf, 38, 39; Jung on leaf, 34, 35; Malpighi on leaf, 38, 39; non-vascular, 58; ontogeny, 79, 107, 108, 178; partial-shoot theory of leaf, 70–123 *et passim*; peltate, Casimir de Candolle on, 109; rachis- or midrib-leaf, 85, 122 (Fig. 22), 123; radiality, 107 *et seq.*; relative growth and leaf anatomy, 170, 171; repetitive branching, 136–9; terminal leaf, 124; transition from foliage leaves to hairs, 139 *et seq.*; vernation, 38 (see also 'Petiolar phyllode'; references, *passim*, to foliage-leaves and other phyllomes)
Leguminosae; bundles replaced by bundle-rings, 104, 105; flower compared with *Polygala*, 155; winged axis, 87; *Amherstia*, 56, 57 (Fig. 5, C), 104–6 (Fig. 14, A); *Amicia*, 81 (Fig. 6, A), 82; *Bauhinia*, 56, 57 (Fig. 5, A), 98 (Fig. 10, A), 99; *Caesalpinia*, 89, 90 (Fig. 8, A); *Cassia*, 56, 57 (Fig. 5, B); *Cercis*, 98 (Fig. 10, B), 99; *Gleditschia*, Frontispiece, B, 98, 120 n., 139; *Lathyrus*, 99, 100 (Fig. 11); *Lupinus*, 111, 113 (Fig. 17, A); *Medicago*, 121 (Fig. 21, D), 169 (Fig. 35, J), 170; *Scorpiurus*, 169 (Fig. 35, K), 170; *Trifolium*, 58, 155, 156 (Fig. 31)
Lesser-celandine, see *Ranunculus Ficaria* L.
Life-knot (*cor plantarum*), 30
Lilac, see *Syringa vulgaris* L.
Lilaea subulata Humb. et Bonpl., anatomy of inflorescence, 94, 95
Liliaceae, *Allium*, 15, 17, 18, 34; *Asparagus*, 97, 193, 194 (Fig. 46, B); *Danae*, 97; 'Lillies', 27; *Ruscus*, 18, 97; *Scilla*, 179 (Fig. 39, C), 180; *Semele*, 97; *Smilax*, 17; *Tulipa*, 44
Lillie, R. S., on teleology and mechanism, 203
'Lillies', Bacon, F., on, 27
Lime tree, see *Tilia europaea* L. and *T. tomentosa* Moench
Linaria vulgaris Mill., peloria, 178, 189 (Fig. 44, C), 190
Linnaean botany, 41

238

INDEX

Myosotis, distribution of blue and yellow coloration, 147

Myosurus minimus L., elongated receptacle, 49 (Fig. 1, B), 50

Narcissus, relation of corona and perianth, 82; Theophrastus on, 143

Narcissus Bulbocodium L., corona dominating parent phyllome, 103 (Fig. 13, E), 104

Nasturtium officinale R.Br., accessory flowers, 178, 179 (Fig. 39, B), 180; stipules, 99, 140 (Fig. 26, E)

Negative conditions, 197, 198

Nelumbium speciosum Willd., Theophrastus on, 21, 22

Neo-Platonism, 209

Nepenthes, pitcher leaves, 157

Nepeta Glechoma Benth., hermaphrodite and female flowers, 192 (Fig. 45, D), 193

Nerium Oleander L., Troll, W., on, 145

Nettle, see *Urtica*

New Holland, Corréa da Serra on vegetation of, 60

'New Philosophy' of the seventeenth century, 37, 202

Nicolaus Damascenus, influence on Albertus Magnus, 24; on self-maintenance in plants, 76

Nicotiana Tabacum L., foliar development, 79; leaf anatomy depending on relative growth, 170, 171

Nœud vital (*cor plantarum*), 29, 30

Norman, J. M., on stipules in Cruciferae, 99 *n.*

Numerical element in biology, 27

"*Nymphaea micrantha*", see *N. stellata* Willd.

Nymphaea stellata Willd., leaf-budding, 107, 115 (Fig. 18, F)

Nymphaeaceae, *Nelumbium*, 21; *Nymphaea*, 107, 115 (Fig. 18, F), 193

Ochlandra setigera Gamble, stamen anatomy, 56

Ocimum Basilicum L., Theophrastus on, 16 *n.*

Oenone, shoot-thalli, 135

Oken, L., on dorsiventrality of leaf, 87; on the relation of leaf and whole plant, 158; on the relation of leaf nervation and shoot branching, 83

Olea europaea L., Theophrastus on the roots, 15

Olive tree, see *Olea europaea* L.

Onion, Jung on, 34; Theophrastus on, 15, 18, 34

Ontogeny and apical development, 80, 162, 164, 166–8, 170, 171, 178

Opuntia, Cesalpino on, 29

Origin of Species (Darwin, C.), 63, 160

Ostrya carpinifolia Scop., Theophrastus on, 17

Ovule, relation of atropous and anatropous forms, 174, 196, 197; shoot-, leaf-, and trichome-interpretations, 131, 142

Ovuliferous scale of conifers, 127–31; Braun's short-shoot theory, 128, 129; Sachs's theory and its variants, including Delpino's theory, 129, 130

Oxalis enneaphilla Cav., compound peltate leaf, 111, 112 (Fig. 16, B)

Oxalis lasiandra Zucc., 111 *n.*

Oxalis lupinifolia Jacq., compound-peltate leaf, 111

Oxalis Ortgiesi Regel, reduction of terminal leaflet-lobe, 98 (Fig. 10, D), 102

Oxeye daisy, see *Chrysanthemum Leucanthemum* L.

Padua, Botanical Garden and Goethe's palm, 42

Paeonia, Jung on, 73

Paeonia Clusii Stearn, foliaceous sepals, 49 (Fig. 1, F), 50

Paeony, see *Paeonia*

Pagella Archeri Schonl., pseudanthium, 148

Palm, Goethe's (*Chamaerops humilis* L.), 42

Palm stem, Theophrastus on anatomy, 16

Palmae, *Chamaerops*, 42; *Phoenix*, 38

Papaver, crowding of young ovules, 175 (Fig. 37, B); Goethe on double form, 44, 45; reduction of median region in carpel, 104

Papaver Argemone L., crowding in young ovary, 175 (Fig. 37, B)

Papaveraceae, *Chelidonium*, 115 (Fig. 18, E), 122, 177; *Glaucium*, 168; *Papaver*, 44, 45, 104, 175 (Fig. 37, B)

INDEX

Tree and herbaceous habit, Takhtajan,
A., on, 108
Trichomes, relation to phyllomes and
shoots, 139–42
Trifolium fragiferum L., comparison of
calyx with corolla of *T. procumbens* L.,
155, 156 (Fig. 31)
Trifolium repens L., foliaceous carpels, 58
Triglochin palustre L., sterile carpels, 58
Triticum, Malpighi on seedling, 38
Troll, W., on Goebel's rejection of
typology, 159; on the carpel, 58; on
his *Gestalt* theory of the flower, 144
et seq.; on the leaf as a *Grundorgan*,
70; on the metamorphosis theory, 68;
on the type concept, 61; on *Zwischenfiedern*, 123
Tropaeolum pentaphyllum Lam., ap-
proximations to peltation, 111, 112
(Fig. 16, A)
Truffle, Theophrastus on, 15
Tulip, see *Tulipa*
Tulipa, Goethe on, 44
Turnip, Cesalpino on, 31
Turpin, P. J. F., on Goethe, 41, 62; on
pen and pencil, 210
Type concepts, 42 *et seq.*, 59–69, 144
et seq.; replaced by parallelism concepts, 86, 159–61

Uittien, H., on relation of leaf nervation
and shoot branching, 83, 102; on
Umbelliferae, pinnate leaf, 119 *n.*,
umbel, 182, 184
Ulmus, distichous. lateral branches, 88
(Fig. 7, E), 89; Malpighi on leaf
development, 38, 39; root-budding,
132, 133 (Fig. 23)
Umbelliferae, *Angelica*, 114, 115 (Fig.
18, B) 116, 177; *Anthriscus*, 168, 169
(Fig. 35, A), 184–6 (Fig. 42);
Bupleurum, 149, 151, 152 (Fig. 29, A)
Carum, 73, 116; *Caucalis*, 169 (Fig.
35, I), 170; *Coriandrum*, 151, 152 (Fig.
29, B); *Daucus*, 149, 150 (Fig. 28),
151, 152 (Fig. 29, C), 182, 183 (Fig.
41), 184, 185; *Eryngium*, 85, 117
(Fig. 19, B), 118, 122 (Fig. 22, B),
139, 140 (Fig. 26, A, B), 141, 168;
Hacquetia, 148; *Heracleum*, 125;
Scandix, 168, 169 (Fig. 35, B)
Unisexuality and sterility in normally
hermaphrodite species, 192 (Fig. 45),
193, 194 (Fig. 46, C)

Units of the plant body, 70–2, 166, 167
et passim.
Urpflanze, Goethe's, 59, 61–4; por-
trayals by Turpin and Schleiden, 62,
63
Urtica, Cesalpino on sterile and fertile
flowers, 29

Vascular supply, effect on form, 195
Vegetative reproduction in phyllomes,
105, 106 (Fig. 14, C), 107, 115 (Fig.
18, F)
Velenovský, J., on pseudo-simple leaves,
98, 99; on relation of leaf nervation
and shoot branching, 83; on terminal
flower of *Rubus thyrsoideus* Wimm.,
182
Venus's-comb, see *Scandix Pecten-
Veneris* L.
Vialleton, L., on 'bush' analogy in
phylogenetics, 66 *n.*
Vienna manuscript of Dioscorides, 27,
123
Virgula parvula of Albertus Magnus, 73
Visual thought in morphology, 210,
211
Voigt, F. S., on distribution of blue and
yellow coloration in flowers and inflorescences, 147
Voltzia, ovuliferous scale, 127, 128

Wallflower, see *Cheiranthus Cheiri* L.
Walton, J., on the cone-scale of a
Permian *Voltzia*, 127, 128
Wardlaw, C. W., on 'Size-Factor',
190 *n.*
Warming, E., on bud and axillant leaf
as a unit, 125
Water-chestnut, see *Trapa natans* L.
Watercress, see *Nasturtium officinale*
R.Br.
Water-hyacinth, see *Eichhornia speciosa*
Kth.
Wettstein, R. von, on the flower as a
reduced inflorescence, 155, 157
Wheat, see *Triticum*
Whewell, W., on the type concept, 67,
68
Whole plant in relation to its parts and
components, 157, 158
Whole-shoot-hood in leaf, urge to, 78,
79, 93–123
Willis, J. C., on root- and shoot-thalli
of Podostemaceae, 134, 135

246